Contents

Cross-curricular links

Chapter	History SoW	Geography SoW	PSHE and Citizenship	Literacy framework/English		Numeracy framework	ICT SoW
1: Lesson 1		Unit 5	Units 6, 1	Y1, Term 1, T12,13,14,15,16, Y1, Term 2, T22	Y1, Term 3, T21 Y2, Term 1, T15		1F
1: Lesson 2		Unit 5	Units 6, 1	KS1 NC Speaking and listening			1C
2		Units 1, 6		Y1, Term 1, S1, 4 Y1, Term 3, S1	KS1 NC Speaking and listening	Y1 handling data Y2 handling data	2E
3		Units 1, 6		Y1, Term 1, T12,13,14,15,16 Y1, Term 2, T22	Y1, Term 3, T21 Y2, Term 1, T15		1A
4		Units 1, 6	Units 9, 6, 1	Y1, Term 1, T12,13,14,15,16 Y1, Term 2, T22	Y1, Term 3, T21 Y2, Term 1, T15		2A
5		Units 1, 6	Units 9, 6, 1, 2	Y1, Term 1, S1, 4	Y1, Term 3, S1		2B
6		Unit 1	Unit 4	Y1, Term 1, S1, 4; T12,13,14,15,16 Y1, Term 2, T22; W8	Y1, Term 3, S1; T21 Y2, Term 1, T15		2A
7		Unit 1	Unit 4	Y1, Term 1, S1, 4	Y1, Term 3, S1		2A, 3A
8	Units 1–3	Unit 1	Unit 6 Unit 1 Unit 5	Y1, Term 1, S1, 4; T12,13,14,15,16 Y1, Term 2, T22; W8, 10 Y1, Term 3, S1; T21	Y2, Term 1, T15 KS1 NC Speaking and listening		1C
9		Unit 2	Unit 4	Y1, Term 1, S1, 4; T12,13,14,15,16 Y1, Term 2, T22; W8, 10 Y1, Term 3, S1; T21	Y2, Term 1, T15 KS1 NC Speaking and listening	Y1 handling data Y2 handling data	2A
10		Unit 2	Unit 4	Y1, Term 1, T12, 13,14,15,16 Y1, Term 2, T22; W8, 10 Y1, Term 3, T21	Y2, Term 1, T15 KS1 NC Speaking and listening		2B, 3A
11		Unit 2	Units 4, 9	Y1, Term 1, T12, 13,14,15,16 Y1, Term 2, T22; W8, 10 Y1, Term 3, T21	Y2, Term 1, T15 KS1 NC Speaking and listening		2A

The local area

David Flint

Curriculum Focus series

Geography

History

Toys Key Stage 1
Famous Events Key Stage 1
Famous People Key Stage 1
Invaders Key Stage 2
Tudors Key Stage 2

Geography

Islands and Seasides Key Stage 1
The Local Area Key Stage 1

Science

Ourselves Key Stage 1
Plants and Animals: Key Stage 1
Materials: Key Stage 1

Published by Hopscotch Educational Publishing Ltd,
Unit 2, The Old Brushworks, 56 Pickwick Road,
Corsham, Wilts SN13 9BX
Tel: 01249 701701

Written by David Flint
Series design by Blade Communications
Illustrated by Martin Cater
Cover illustration by Susan Hutchison
Printed by Clintplan, Southam

ISBN 1-904307-50-7

Introduction

Curriculum Focus: The Local Area helps to make geography both fun and real by giving teachers (especially those who may not have much geographical background) the material and support that they need to plan and teach exciting and interesting lessons. The chapters are based on the QCA exemplar scheme of work for geography at Key Stage 1 and each chapter equips teachers with ideas, skills and knowledge to deliver the full range of geography at this key stage.

This book gives teachers a clear approach to teaching geographical ideas and to planning work for their classes, including:

- background information that includes illustrations;
- ideas for introducing and developing the lessons;
- differentiated photocopiable activity sheets to support individual and group work.

The local area is the most familiar part of a child's environment; hence it is a logical starting point for geography. The intention for using the local environment is to encourage children to look afresh at their local surroundings and to see and understand things that they may have taken for granted. This also provides an ideal opportunity to develop a range of ideas and skills, some that are unique to geography (such as map skills) and others that are general educational skills (such as description and analysis).

In investigating the local area it is important to help children think about the journeys that they make within that environment. Hence Chapters 3 and 4, together with Chapter 5, help to develop ideas of travel and movement. They also help to develop children's skills in understanding and drawing maps. This is an aspect of geography that has not been made very explicit within the QCA units of work, so the *Curriculum Focus: Geography* series makes a point of showing exactly how teachers can develop map skills with very young children.

Chapter 5 focuses on getting children to think about nice and nasty places in their local area. In this way children can begin to express their opinions about their environment and hopefully to suggest ways in which it can be improved. Chapters 6 and 7 extend the study of the local area to local jobs and local leisure opportunities. Chapter 8 develops the idea that the local environment is never stationary and that change is the norm. Chapters 9, 10 and 11 deal with aspects of roads, traffic, parking and safety. The emphasis is on helping children to think about dangers in the environment and ways of overcoming those dangers.

Curriculum Focus: The Local Area recognises that there will be different levels of attainment among the children and that their developing reading skills will require different levels of support during individual and group work. To help teachers to provide activities that meet the needs of their class, each chapter contains three photocopiable sheets based on the same materials but written for children with different levels of attainment. Activity sheet 1 in each chapter is intended for lower-attaining children. Activity sheet 2 should be suitable for most children while Activity sheet 3 challenges higher-attaining children.

Mapping skills

'History is about chaps and geography is about maps,' is an old saying among teachers that still has a great deal of validity. Children meet and see maps and plans in their everyday lives, from the plan of the play area showing where to find swings, slide and climbing frame, to the map of the supermarket showing where to find frozen foods, fresh meat and vegetables. We often assume that children see the same things on a map or plan as adults see, but this is often not the case. So one of the main purposes of geography is to help them learn to 'read' maps in the same way as they develop literacy skills in reading text. In fact, the skill of using, making and reading maps, called 'graphicacy' or 'visual literacy', ranks alongside traditional literacy and numeracy in terms of children's development. However, graphicacy has never been afforded the importance and time that have been devoted to literacy and numeracy. One of the purposes of this series of *Curriculum Focus* books is to help to rectify this neglect.

Children need to be taught how to read a map. It is not a particularly difficult skill and it can involve games and fun activities. One of the problems with the teaching of map skills is that they have to be taught by teachers who are probably not geographers themselves and who may be worried about their own map reading abilities. If this is you, please don't worry – teaching map skills is simple, enjoyable and fun for teachers and for children. These books will show you how to do it in easy steps.

The QCA units of work in geography form the basis of these books. These units of work are extremely useful with their detailed learning objectives and suggested teaching activities. However, many of the units contain references to the development of map skills, such as 'Locate Scotland and the Western Isles on a map,' (Unit 3, An Island Home), without showing how these skills are to be acquired. These books will help to fill this gap, starting with Chapter 1 on how to build up a series of activities to develop map skills.

Maps and mapping

TEACHERS' NOTES: LESSON 1

This book starts with this chapter on ways to develop map skills because they lie at the heart of geography. The intention is to help you and the children to get a flying start in understanding geography and how it works. You may want to refer back to the two lessons in this chapter from time to time as you work through other chapters to remind the children of some of the skills they have acquired. This will help them when they come to use or draw maps in later chapters, such as the ideas about plan view in Chapter 10. You will find yourself revisiting Chapter 1 as you work with the children throughout the year.

Work on developing map skills should permeate all aspects of geography, so that children see maps as an integral part of any study, whether the place be local or distant. The development of their map skills is an ongoing process that starts in nursery and continues throughout their lives.

Some basic rules about developing map skills

First of all, there are two big don'ts:

- Please don't start the first lesson on map skills with 'Today we are studying geography and so I want you all to draw a map of your route from home to school.'
- And please don't start the next lesson with 'Next we are going to draw a plan of the classroom.'

These are very difficult activities for adults, let alone children. It is really important that children do not find that they fail in their first activity related to geography. The development of map skills should be fun, enjoyable and, above all, easy. It is important to differentiate the activities so that all the children can see that they are succeeding in map reading from the very first lesson. One interesting finding from recent research is that children who may not be very successful in traditional literacy and numeracy can be extremely successful in terms of map skills, and this may help to motivate all the children in the group.

So the basic rules for developing map skills are as follows:

- Keep it simple.
- Make it fun.
- Use lots of games.
- Avoid overly-detailed and complex maps.
- Build in success.
- Use lots of picture maps in the early stages.
- Avoid complex terminology in the early stages.

The 'Curriculum Focus' approach to the development of map skills

With all of the above in mind, we have drawn on a wide range of experience in teaching map skills to produce a guide to how best to develop children's map skills. What follows is a suggested series of activities after which children will be able to draw, use and understand maps. This sequence of activities can be used with young children, but can also be adapted for older children. (The sequence of activities remains the same but older or more able children will be able to work through the activities more quickly.) Remember that the key to all this is success. Children need to see themselves as successful in the development of their map skills.

Activities for the development of map skills

The key elements needed to draw, use and understand maps are:

- the language of location (using words to describe where things are);
- directions – from left and right to north, south, east and west;
- understanding and using signs and symbols;
- understanding the idea of plan view.

At Key Stage 2 these elements continue and become more complex, and two further elements are added:

- coordinates;
- scale.

Lesson 1 will take you as far as location and directions. Lesson 2 develops understanding and the use of signs, symbols, and plan view.

Maps and mapping

Geography objectives
- To learn some of the language of location.
- To learn the four and eight points of the compass.
- To draw a picture map.

Resources

- Generic sheets 1–5 (pages 10–14)
- Activity sheets 1–3 (pages 15–17)
- Blank A4 paper
- Pencils, pens, rubbers
- Marker pens

Starting points: *whole class*

Tell the children that they are going to draw a picture of a park showing where things are to be found. Explain that they have to listen very carefully to your instructions about what to draw and then draw it very carefully. Read through the instructions on Generic sheet 1, a line at a time, pausing to allow the children to draw the feature before moving on to the next line.

As the children finish the picture tell them that this is a picture map. It shows where things are, so it is a map. Stress the words you have used to describe location, such as 'next to', 'at the bottom', 'behind' and 'from the front'.

Now show the children Generic sheet 2 on an OHP or in an enlarged version. Explain that this illustrates some of the words we use to describe the position of things. Key words are:

- on
- off
- inside
- outside
- in front of
- behind
- next to
- far away

Ask a child to come out and to draw a line connecting pictures that are opposites, such as 'in front of the picture' and 'behind the picture'. Repeat this until all the pictures are linked. Explain that these words and pictures help us to find where places are.

Next show the children Generic sheet 3 on an OHP or in an enlarged version. Explain that Shifali wants to reach her friend Sally and has to find a path through the maze. Ask a child to come out and to draw a line through the maze from Shifali to Sally. Ask another child to put a circle every time Shifali turns left in the maze and a cross every time Shifali turns right. The aim here is to help the children to project themselves into the map of the maze and to be able to think about which directions they turn in order to escape from the maze. Ask 'How many times did Shifali turn left?' and 'How many times did she turn right?' Explain that 'left' and 'right' are directions.

Now show the children Generic sheet 4 on an OHP or in an enlarged version. Explain that we use a compass to describe directions. Point out north, south, east and west and the use of single capital letters for these four points. Ask 'What is north of Sophie?', 'What is south of Sophie?', 'What is west of Sophie?' and 'What is east of Sophie?' Explain that we find things on maps by using directions.

Next show the children Generic sheet 5 on an OHP or in an enlarged version. Explain that we can give more accurate directions using the eight points of the compass: N, NE, E, SE, S, SW, W, NW. Ask 'Which animal is to the north-west?' (monkey), 'In which direction would I go to find giraffes?' (SW) and 'In which direction would I find the lions?' (N)

Tell the children that they are now going to draw a picture map using directions.

Group activities

Activity sheet 1
This sheet is aimed at children who need more support. They are able to identify the four main points of the compass. They have to complete the picture by drawing in a house, a lake, some tall

trees and a car park around the hotel. If there is time, they could add other features to the picture using the correct directions, such as a car to the north and a bus to the east.

Activity sheet 2
This sheet is aimed at children who can work independently. They are able to identify the eight main points of the compass. They have to complete the picture by drawing in features at the correct compass points around the hotel. If there is time, they could add other features, such as a river to the east of the hotel and a road to the south.

Activity sheet 3
This sheet is aimed at more able children. They are very familiar with the eight main points of the compass. They have to complete the picture by drawing in features at the correct compass points around the hotel.

Plenary session

Share some of the responses to the activity sheets. Recap both the four and eight main points of the compass.

Talk about the picture maps and what they show in the area around the hotel (what the main features of the area look like and where they can be found).

Talk about some of the problems with these picture maps, such as the difficulty in fitting in a picture of all the features. Ask the children for alternative ways to show these on the map. Suggest the idea of a symbol for each feature, which would take up less space and still tell people what it was.

Ideas for support

To help children learn the idea of the order in which 'north', 'south', 'east' and 'west' come, remind them of simple mnemonics such as Never (N) Eat (E) Shredded (S) Wheat (W).

To help children grasp the importance of words that describe location, play games such as 'Simon says'. ('Simon says stand BEHIND the door,' 'Simon says stand ON the chair,' 'Simon says stand NEAR TO the window,' and 'Simon says stand UNDER the light.') They will soon see that words that describe location are really useful and easy to use. Similarly with directions, start with 'left' and 'right' and play trails around the classroom, such as 'Stand with your back to the door. Take two paces forwards. Now turn right. Take one pace forwards. Now turn left. What is in front of you?' These simple games can end with a person or a classroom feature such as the whiteboard or the window.

Ideas for extension

Ask the children to draw a signpost map of the children sitting around them. They should draw the eight points of the compass and then write the name of the child who is sitting N of them, NW of them, W of them and so on to include SW, S, SE, E and NE. If there is no child in one of the directions, ask them to write this on the signpost map.

Let the children see a compass and use it to find N, S, E and W around the school. Then ask them to draw pictures of what they can see to the N, S, E and W of their classroom.

Linked ICT activities

Using remote-controlled toys, place a set of two or three obstacles on the floor and ask the children to move their toys around the obstacles using simple instructions and directions to get the toy from A to B. Encourage them to use directional language, and give them the opportunity to talk about what they are doing – for example, 'I am going to move forwards and then turn.'

Progress to using either a Roamer or a Pixie, which are generally used within the Key Stage 1 setting, and allow the children to input more direct instructions. Place A4 sheets of card on the floor and label them with the names of the different buildings which they may find in their town/village or the different things that they may find in the park. Place the obstacles on the floor, giving plenty of room to move the Roamer/Pixie around the obstacles. Set a simple task for the children – for example, the Roamer needs to move from the play area to feed the ducks on the pond. Place the Roamer at the play area and talk to the children about each instruction they are going to give the Roamer to move it to the duck pond.

Maps

Draw a pond as big as your little finger in the middle of your paper.

Draw a tall tree next to the pond.

Draw two birds sitting in the tree.

Draw some tall grass at the bottom of the tree.

Draw a low fence behind the pond and behind the tree.

Behind the fence draw a house with three windows and a door.

Draw a footpath from the front of the picture to the pond.

Draw a boy with a dog on a lead walking on the footpath.

Draw a flowerbed somewhere.

Draw four clouds in the sky.

Maps

Maps

Maps

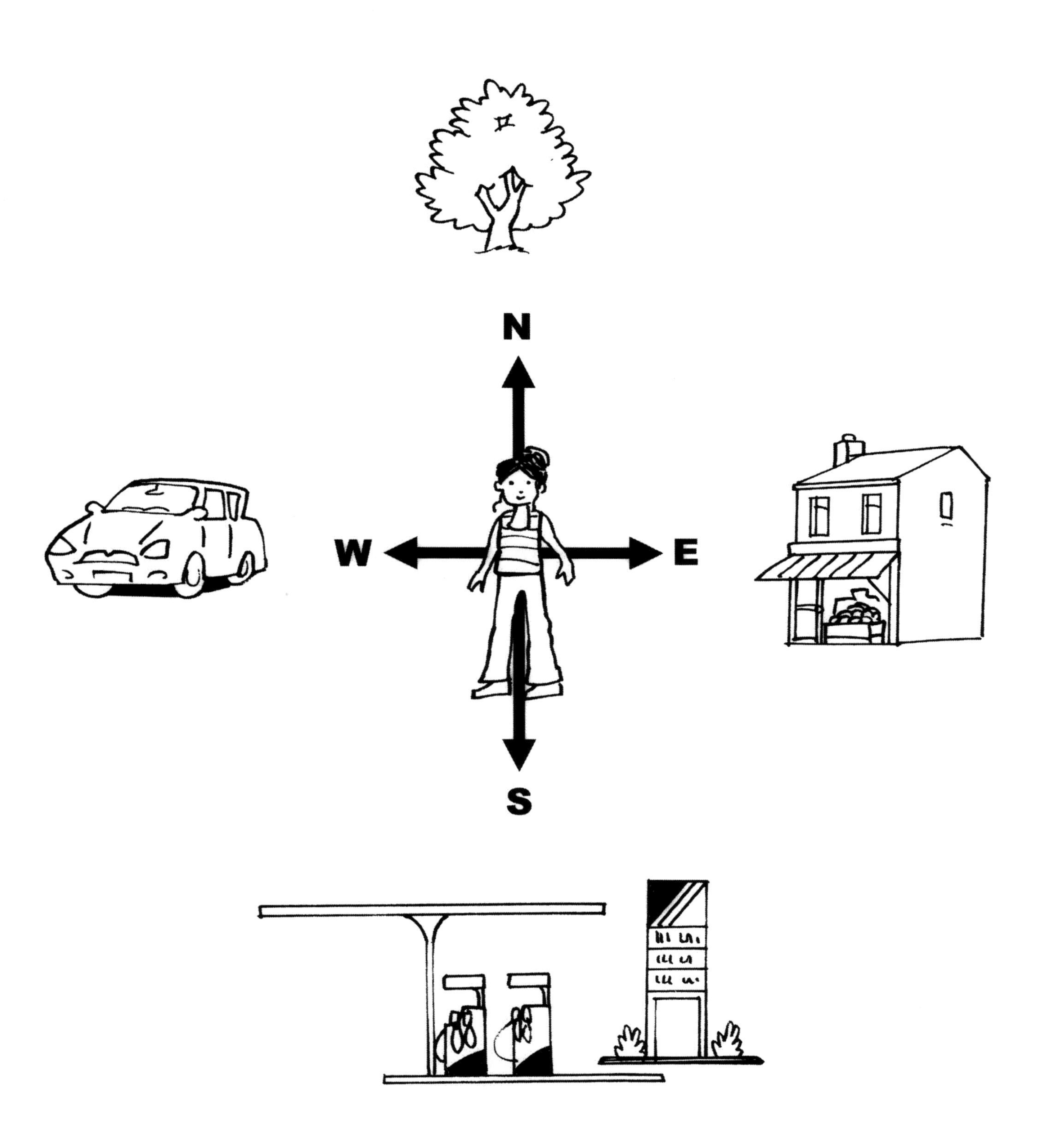

Maps

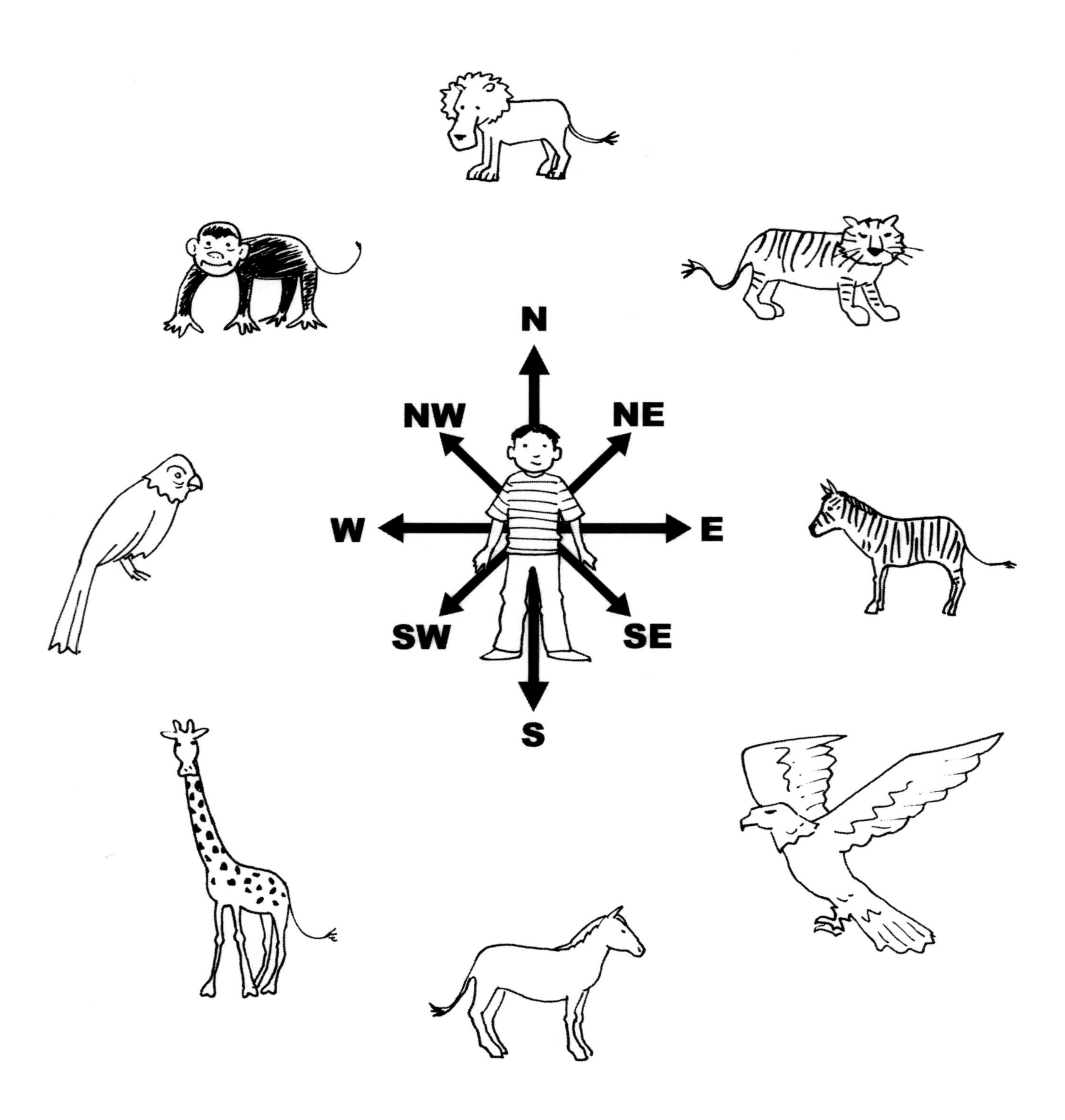

Name ______________________________

Maps

Complete the picture map by drawing these things in the correct places:

Draw a house west of the hotel.
Draw a large lake east of the hotel.
Draw five tall trees north of the hotel.
Draw a car park to the south of the hotel.

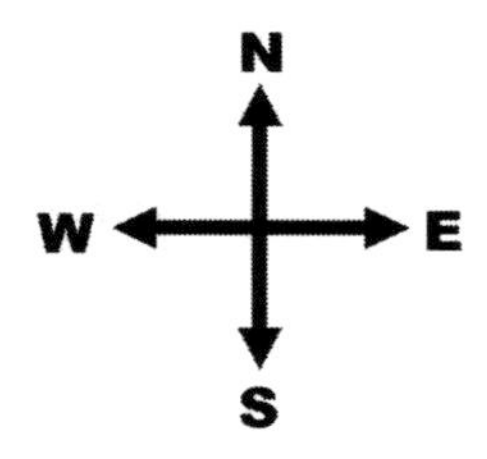

Name ______________________________

Maps

Complete the picture map by drawing these things in the correct places:

Draw a large lake with boats on it to the south-east of the hotel.
Draw a forest of trees to the north-east of the hotel.
Draw a mountain north of the hotel.
Draw a car park south-west of the hotel.
Draw a church to the north-west of the hotel.

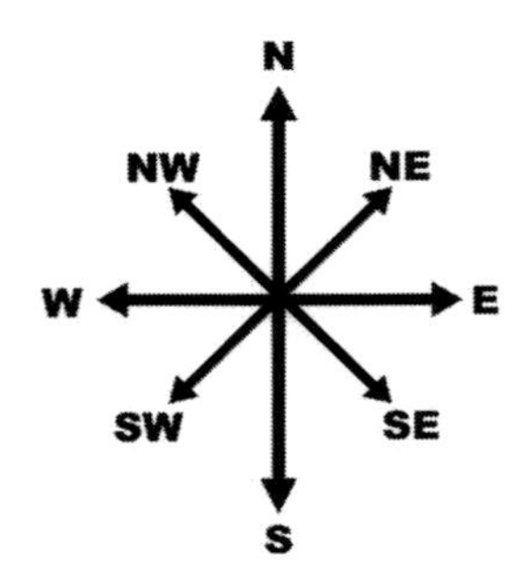

Name ______________________________

Maps

Complete the picture map by drawing these things in the correct places:

Draw a large lake with boats on it to the south-east of the hotel.
Draw a forest of trees to the north-east of the hotel.
Draw a mountain north of the hotel.
Draw a car park south-west of the hotel.
Draw a church to the north-west of the hotel.
Draw a river entering the lake to the east of the hotel.
Draw a windmill to the west of the hotel.
Draw a busy main road to the south of the hotel.

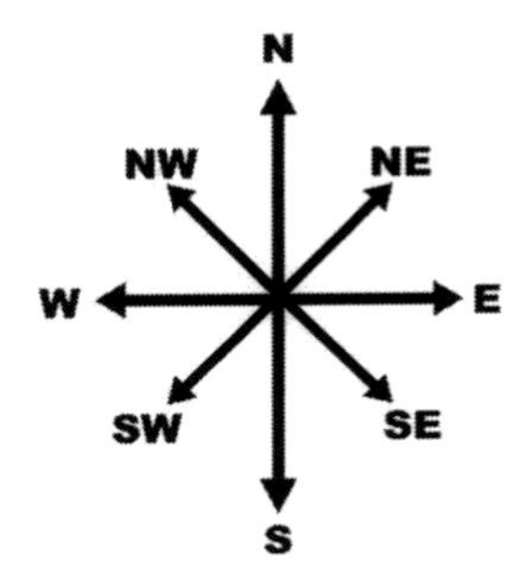

Maps and mapping

TEACHERS' NOTES: LESSON 2

Lesson 1 showed you how to start developing children's map skills and how to make it fun and enjoyable. The *Curriculum Focus* approach to the development of map skills is based on a series of activities that enable children to draw, use and understand maps. This lesson continues the sequence of activities begun in Lesson 1 to enable all children to draw and use maps with confidence. The additional elements are:

- understanding and using signs and symbols;
- understanding the idea of plan view.

Understanding and using signs and symbols

Children are used to seeing and using signs and symbols in their everyday life. They see road signs on their way to and from school. Many road signs contain no words, but use pictures, a diagram or shapes to convey their meaning. Circular signs with a broad red line diagonally across them mean 'do not do' something, such as 'Do not turn right.' Triangular signs advise, such as 'Give way.' Circular road signs with a red band round have to be obeyed, such as speed limit signs. Rectangular signs provide information, such as 'One way street.'

It is important to help children understand that in the same way, signs and symbols used on a map convey information – a large letter T shows the location of a telephone box (although some might have a telephone handset symbol, in different colours depending on the type, such as black for public and blue for an AA phone box); a square with a cross on top shows a church with a tower. The children do not need to adopt these conventions at the outset. Encourage them to devise and draw their own signs, such as signs for different types of shop – some selling meat, or vegetables, or newspapers. The important points to stress are:

- that the sign looks like what it represents, such as a loaf of bread for a baker's;
- that there is a key, which explains what each sign stands for.

Tell them that the word 'key' is used deliberately because it is the thing that unlocks the whole of the map.

Plan view

This can be one of the more difficult aspects of maps for children to understand. The key elements are:

- things (objects) look different when we look down on them from above;
- the view from above gives us clues as to what objects are by their shape;
- the view from above makes it hard to see how tall the object is.

The main point is not to expect too much from the children too soon. Introduce them to the idea of plan view gradually, through a series of games and activities. For example, put a variety of objects on the OHP (but hidden from the children) and asking them what they think the objects are; make some cards of objects seen from above and the front and ask the children to match them. Objects for the above two games could include a lamp, a kettle, a cup, a pen and so on.

Maps and mapping

Geography objectives
- To learn how to draw and interpret signs and symbols on maps and plans.
- To identify objects from their plan view.

Resources

- Generic sheets 1–5 (pages 21–25)
- Activity sheets 1–3 (pages 26–28)
- A set of photos of toys taken from directly above
- Coloured marker pens
- Blank A4 paper
- Pencils, pens and rubbers

Starting points: *whole class*

The activities suggested here will probably have to be done over a series of sessions. Tell the children that they are going to look at some of the signs and symbols used around us and also used on maps. Show them Generic sheet 1 on an OHP or in an enlarged version. Explain that in a large supermarket it is important to help people to find out where things are, so signs and symbols are used on the tops of shelves or hanging from the ceiling. Ask:

- What would we find in a supermarket near signs 1 to 4? (Fish counter, fresh meat, bread and cash machine.)
- What would we find near signs 5 to 6? (Washing powder, toothpaste, frozen food and fizzy drinks.)

Then ask what could be drawn on sign 7 to represent frozen food. Ideas might include ice cubes or the words 'Frozen Food' or 'Freezer', but encourage them to suggest other ideas as well. Similarly ask for ideas for fizzy drinks on sign 8.

Next show the children Generic sheet 2 on an OHP or in an enlarged version. Explain that we use lots of different symbols like these. Point out that we can use the symbols to make messages, as in the first sentence, 'I will meet you by the tree.' Ask the children to say what each sentence means in turn. (Sentence 2 is 'Are you happy?'; Sentence 3 is 'The road is busy.') Now ask them to write their own messages using as many symbols as possible. Compare a few of these messages.

Next show the children Generic sheet 3 on an OHP or in an enlarged version. Tell them that they are now going to think about the plan view of things, because maps all use the plan view. Explain that things look different when seen from directly above in the plan view. Ask a child to come out and draw a line between the picture of the phone box and its plan. Point out that we can identify it by its shape. Now do the same with the tent, asking another child to link the picture with the plan. Repeat this for the bus, wheelbarrow and cup.

Now show the children the picture and the plan of Jo's living room on Generic sheet 4 on an OHP or in an enlarged version. Explain that these show the same room – the 'plan' shows how the room looks in plan view. Ask a child to come out and colour the cat black on both the picture and the plan. Next ask a child to come out and colour the sofa green on both picture and plan. Repeat this with each item on the sheet: armchair (blue), table (yellow), television (brown) and fireplace (red).

Tell the children that they are now going to look at some more examples of objects in both picture and plan view.

Group activities

Activity sheet 1
This sheet is aimed at children who need more support. They are able to identify the picture of the sandpit but may need some help in identifying the plan view of each object. They have to match each toy in the picture with the plan view and colour both in the same colour. On the back of the sheet they have to draw the plan view of a teapot.

Activity sheet 2
This sheet is aimed at children who can work independently. They are able to identify the picture and plan view of objects quite easily. They have to match three toys in the picture with the plan view and colour them in the same colour. They also have

to identify the toys missing from the plan and draw these in the correct places and in the correct plan shapes.

Activity sheet 3
This sheet is aimed at more able children. They can easily identify the picture and plan view of objects. They have to match the picture with its plan view and shade both in the same colour. Then they have to identify the four missing toys and draw each on the plan in its correct place and in the correct shape.

Plenary session

Share the responses to the activity sheets. Refer back to examples of signs. Ask:

- What sign could be used to show a school (or a library, or a hospital) on a map?
- What clues did you use to identify the plan views of the toys? (Shape and size.)
- Why is it sometimes difficult to recognise toys in the plan view? (Because they look different from the side view that we normally see.)

Use Generic sheet 5 on an OHP or in an enlarged version to show how houses – detached, terraced, blocks of flats – look different when seen in plan view. Stress the fact that we cannot see how high the block of flats is. Point out that maps use the plan view.

Show photos of children's toys taken from directly above the toy. Ask the children to look at the photos and to find the correct toy from a box in the classroom. Point out how the photo is different from the reality and point to ideas for how to recognise objects by the clues from their shapes.

Ideas for support

To reinforce the idea of 'shape' clues, use the OHP and explain that the light shows objects seen from directly above. Put a series of objects on the OHP such as a pen, a small pair of scissors and some keys. Ask the children to guess what each object is and ask what clue they used (usually this is shape). Next put a series of coins on the OHP such as 1p, 5p, 10p, 20p, 50p, £1 and £2. Again, ask the children to try to recognise each coin from its shape on the OHP. Some, such as the 20p and 50p, are relatively easy to recognise. However, size is another clue to distinguish 1p from 5p and 10p and £1. Point out how the plan view makes it hard to judge the thickness of objects such as coins, and in the same way it is hard to judge the height of houses in pictures taken from directly above.

Ideas for extension

Give the children a plan view of the classroom that you have drawn, showing the location of the tables, chairs and the door. Ask them to colour the chairs in one colour and the tables in a different colour. They could then add plan views of other items in the room, such as the mat, and colour that in. Finally they should add a key below the plan to show what each colour represents.

Linked ICT activities

Work with the children using a child-friendly digital camera. Choose ten objects that they are familiar with from around the school. Developing the concepts about viewing objects from above, ask them which of the ten objects they think will be the easiest to identify from photographs taken above the objects. Ask them what clues they think will help them to identify these objects. Take a photograph of one of the objects from above and display it on the computer. Show the children both the object and the photograph. Which parts of the object are clearly recognisable from the view above? If the object was a kettle, they would clearly be able to see the shape of the spout. Help them to look carefully at the clues in the picture. Then ask them to choose another object that they will clearly recognise from a photograph taken above the object. Were they right? As you move through each object, the children will become more familiar with the type of clues they are looking for in the picture. Give them the opportunity to use the camera to take a 'bird's eye' picture.

Show the children other photographs of objects that may not be as easy to recognise. Are there any clues in the picture to help them?

Maps and mapping

Maps and mapping

Here are some symbols and their meanings:

	yes		no
	sunny		cloudy
	happy		sad
	tree		lake
	railway line		road

Write what the symbols mean in these messages:

I will meet you by the . ________________

Are you ? ________________

The is busy. ________________

Use the symbols to write a message of your own.

Maps and mapping

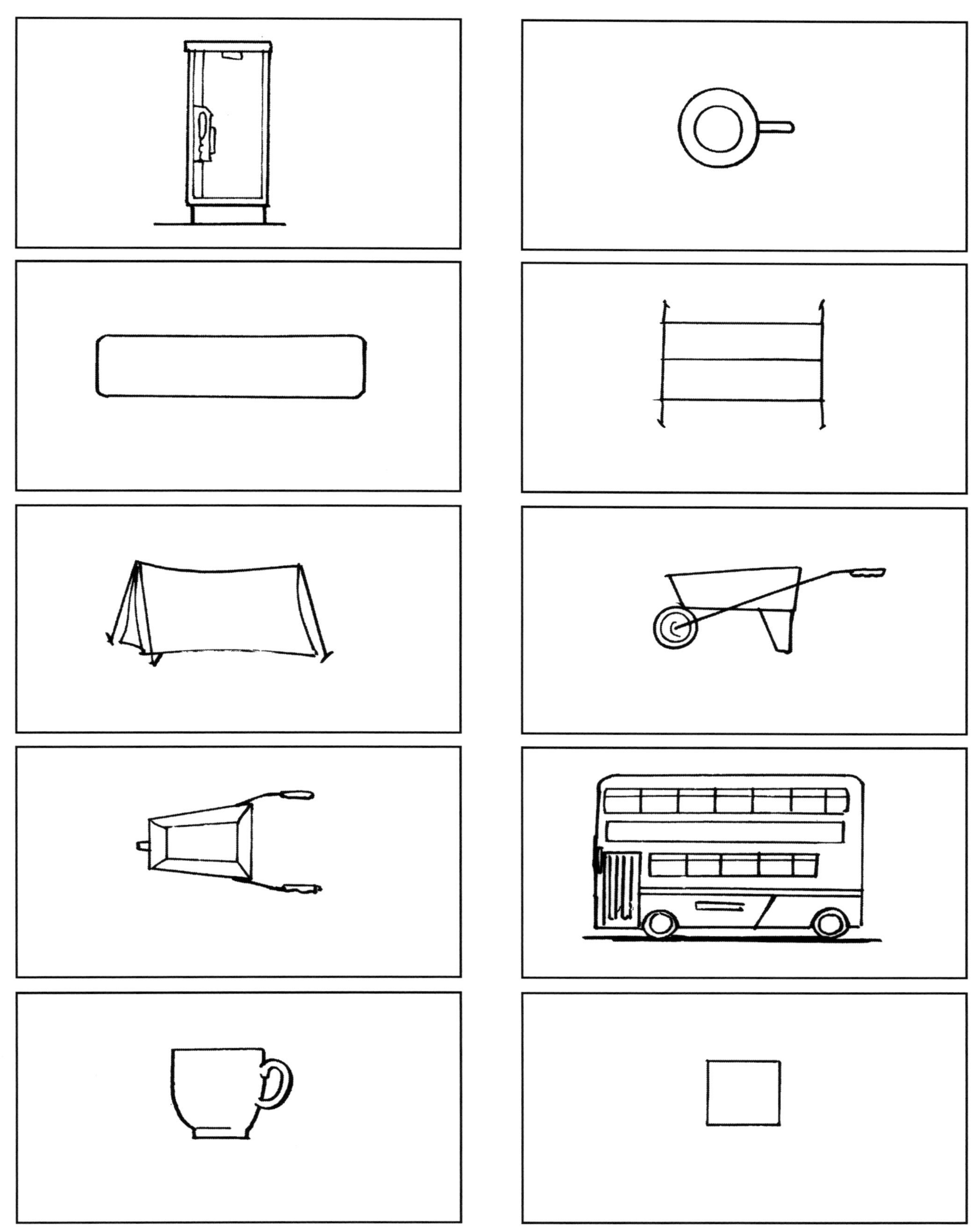

Maps and mapping

This is Jo's living room.

This is a plan of Jo's living room.

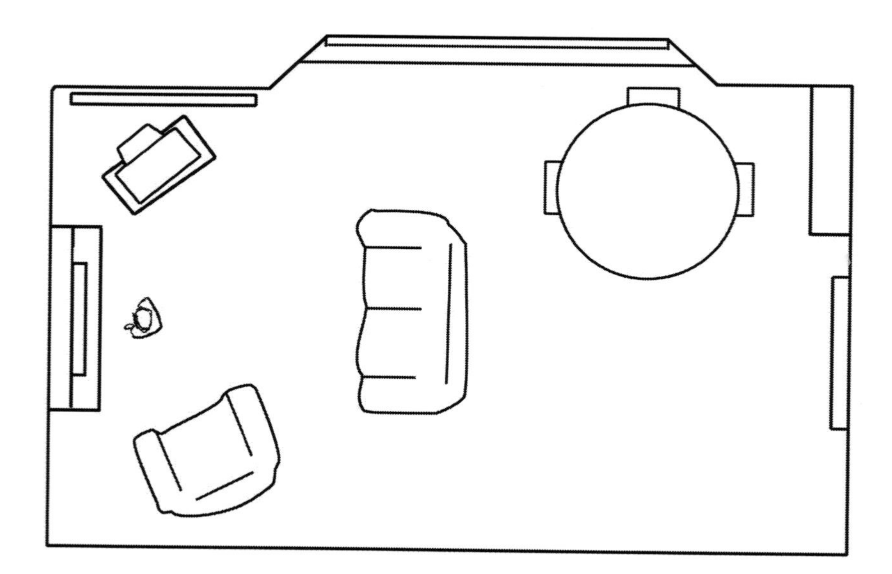

Colour these things in the picture and on the plan:

cat (black)	television (brown)	armchair (blue)
fireplace (red)	table (yellow)	sofa (green)

Maps and mapping

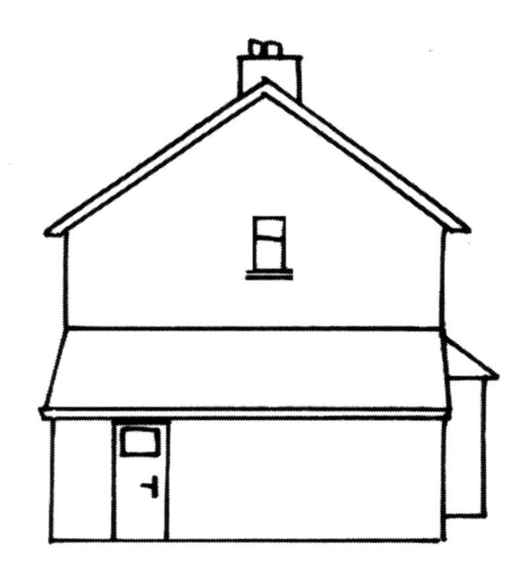

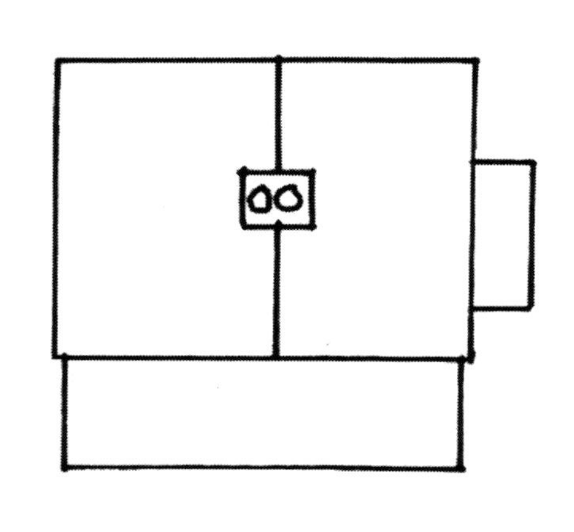

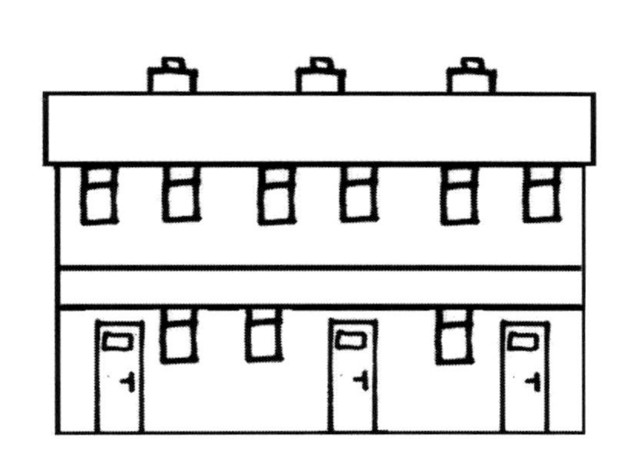

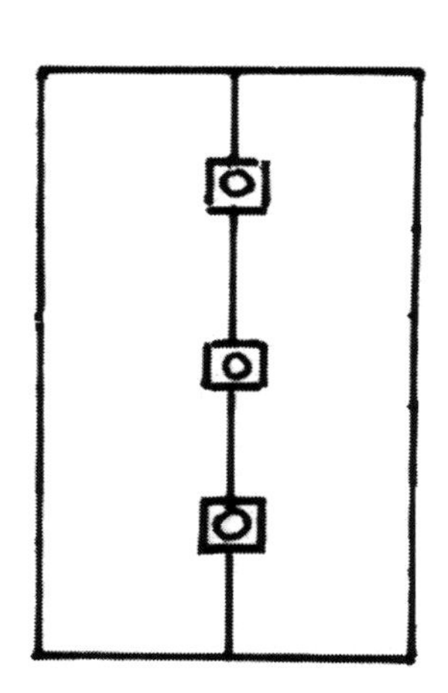

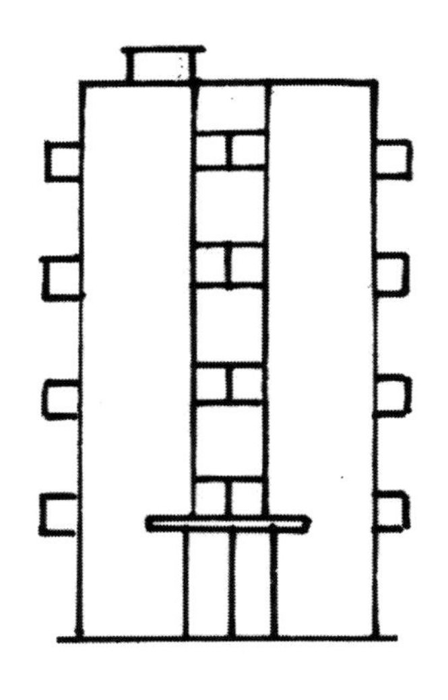

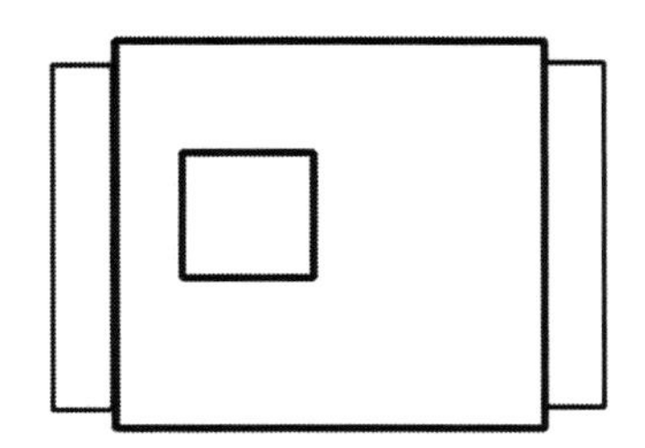

Name ______________________________

Maps

This is a sandpit with some children's toys.

Below is a plan of the sandpit. Match each toy by colouring each picture and its plan view in the same colour.

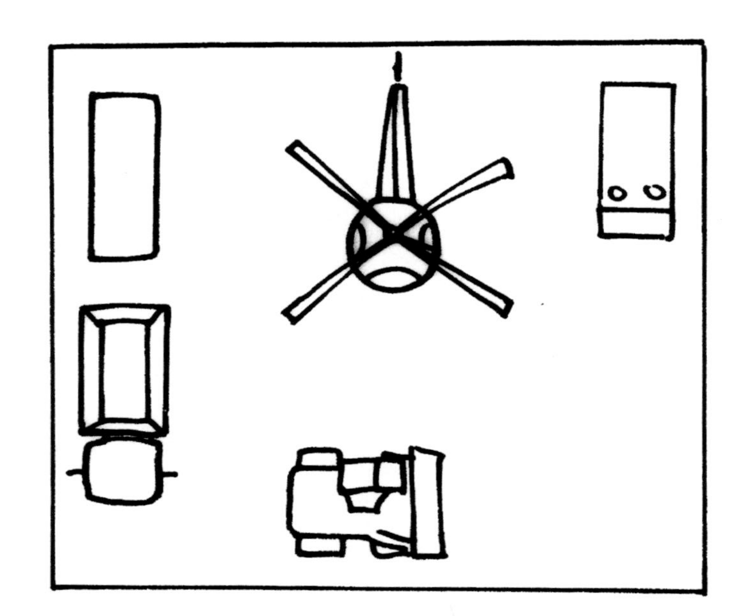

On the back of this sheet, draw a plan view of a teapot.

Name ______________________________

Maps

This is a sandpit with some children's toys.

Below is a plan of the sandpit. Match the three toys by colouring each picture and its plan view in the same colour.

Two toys are missing from the plan. Draw in the plan view of each missing toy.

Name ________________________________

Maps and mapping

This is a sandpit with some children's toys.

Below is a plan of the sandpit. Match the toy by colouring its picture and its plan view in the same colour.

Four toys are missing from the plan. Draw in the plan view of each missing toy.

Where I live

CHAPTER 2

TEACHERS' NOTES

This chapter highlights the components of an address and develops ideas about the importance of individual addresses. It also deals with approaches to studying the different ways in which people travel to school.

Addresses

Addresses are important because they are the means by which we receive goods and services. The delivery of letters and parcels is the most obvious example of this, but we also have 'deliveries' of water, gas, electricity, telephone and other services. The costs of these services form a significant proportion of household budgets, so it is important that everyone is charged only for those services they have received. This is largely achieved through an address.

There are a number of issues relating to individual addresses to which you need to be sensitive:

- Not all children, either in the class or in the area, will have an address. Some children may be staying with relatives and not want to reveal this. For a variety of reasons, children may be staying in hotels, hostels or bed and breakfast accommodation and again this needs to be handled sensitively. Children in travelling families may not have a traditional address. In all these cases it may, therefore, be better to use the address of the school. Visits to the playground to examine nearby street names can be used to confirm some of the elements of an address, together with the school headed notepaper.

- In rural areas, street names may be very infrequent or non-existent. Some houses or farms may be simply known by their name and the village or area where they are. So it is important to emphasise that not all addresses start with a house or flat number or name.

- Many children live in blocks of flats. It is important not to imply that this is in any way inferior to living in a house. It is equally important to avoid qualitative judgements that might suggest that living in a detached house might be better than living in a semi-detached or terraced house. (Many flat dwellers have such astounding views across the surrounding area that they may not want to live in a low-rise, high-density housing development.)

House numbering
In some streets, houses are numbered consecutively along the road. But in many cases, houses are numbered on an odd and even basis with all the odd numbers on one side of the road and the evens on the other. It is important to point out to children that postal workers need to be aware of these variations in order to deliver the post as efficiently as possible. In most cases, delivery workers sort the letters into streets, and then sub-divide them on the basis of the numbering system. Then the post for all the even numbers along a road will be in one bundle and that for odd numbers in another.

Flat numbering
It may be useful to point out to children the ways in which flats are numbered. This usually starts from the bottom and works up, for example 1 to 65, or flats may be numbered in relation to the floor they are on: so flats 110 and 115 would both be on the first floor, and so on.

The concept of an address is a difficult one for some young children to grasp. They are aware of the house or flat where they live and recognise it by features such as the colour of the door or the level of the letter box. They recognise the homes of friends and relatives by such features, or even by the people who open the door. Young children are used to being moved from one place to another. This movement may be by car or bus, which adds to the feeling of disorientation. So when the car or bus arrives, the child might have little or no idea where they are or how far they have travelled. Still less might they know the name of the street they have come to, or the number of the house.

Young children can be taught to recite their address and this may be very useful. However, this rote learning needs to be reinforced with walks along the road to identify street names and house numbers. In this way they begin to associate the rote learning with reality.

Where I live

Geography objectives (Unit 1)
- To know that most pupils have a personal address and that they travel to school.

Resources

- Generic sheets 1 and 2 (pages 32 and 33)
- Activity sheets 1–3 (pages 34–36)

Starting points: *whole class*

Tell the children that they are going to look at addresses where people live. Show them Generic sheet 1 on an OHP or in an enlarged version. Explain that the top picture shows Joanne and her mum on the way to school. The bottom picture shows terraced houses and Joanne lives in the middle one. Ask questions about the house, such as:

- How does Joanne get in?
- How many windows are there?
- What are the windows made from?
- What is on the chimney?

Now show the children Generic sheet 2 with Joanne's address on it. Take them through it, line by line, asking questions such as:

- What number is Joanne's house?
- Why does her house have a number?
- What street does Joanne live in?
- What part of town does Joanne live in?
- Which city does Joanne live in?
- What do the letters and numbers mean on the envelope?

Ask the children to look again at the picture of Joanne and her house. Ask:

- How does Joanne get to school?
- What other ways might she get to school?

Tell the children that they are now going to think about where they live, and how they travel to school. They will be thinking about their home and their address.

Group activities

Activity sheet 1
This sheet is aimed at children who need more support. First they have to draw a picture of their house and then a picture of themselves on their way to school to show how they travel. Finally, they have to write their name and address. Some of them may need to have their address written out for them to copy.

Activity sheet 2
This sheet is aimed at children who can work independently. They are familiar with their address and its different parts. They have to fill in their address. Then they have to draw a picture of themselves travelling to school. On the back of the sheet, they have to draw their home.

Activity sheet 3
This sheet is aimed at more able children who are familiar with their whole address and are able to write it in full. Once they have done this they have to identify their method of travel to school and say how long the journey takes. They have to identify the type of home they live in. On the back of the sheet, they have to draw their home.

Plenary session

Share some of the responses to the activity sheets. Recap the main components of the address – house number, street name, area, town and postcode.

Take a census of how many children travel to school on foot, by car, on the bus, by bicycle or by other means. Find out which is the most popular method of travel. Ask questions such as 'Do the children who come by car live the furthest from school?' and 'Who has the longest journey?'

Ideas for support

To help children grasp the importance of addresses, make up a large parcel with the name of their class and the school address on it. Go over the parts of the address – name of school, street name, area, town and postcode. Then make up lots of small parcels and ask each child to write their name on one of them. Explain that if they now go to the post office and put the parcels in the post with just a name, the postal workers will not know where and how to find where to send them. Ask them what else the postal workers will need to know in order to deliver the parcel (class, school, street, area, town and postcode).

Ideally, post a letter to each child at their home address, so that they can bring them in to show each other.

Ideas for extension

Ask the children to use the responses from the activity sheets to draw a picture graph of the number of children getting to school by car, bus, bicycle, walking or other means.

Mark the different addresses of children in the class with coloured circles on a large-scale map of the area around the school. Each circle can be named for a child, or a small photo of each child can go inside the circle. Then ask the children to decide who has the furthest to travel to school and who the shortest distance.

Discuss postcodes with the children. You will need a map of local postcodes (available from the local post office). Ask them what the first letters mean (an 'area code' relating to the nearest large city, or geographical area. Birmingham, Glasgow, Liverpool and Sheffield have single-letter area codes). London has single and double, such as W (West) and SW (south West). Other places have two letters (for example CV: Coventry and PR: preston). Ask the children what the numbers in the first part of postcode mean (the 'district', where 1 indicates the centre of that city, with 2 onwards indicating how far away you are). The second part of the postcode has a number to show a 'sector' within your district. The final two letters are unique to a group of houses or an individual building. (So, in theory, the postal worker just needs your postcode and the number of your house.)

Linked ICT activities

Having gathered the information from the children about what type of house they live in and how they travel to school, this can be used to create an information handling activity using the computer.

Using Activity sheet 1 as a starting point, create a visual display of how the children get to school. Use the sheets that the children have completed and pin them on the wall under the headings 'We walk to school', 'We come by bus', 'We come by car' and 'We come by taxi'. Talk to the children about the pictures. Can they count how many children walk to school, and so on?

Use a simple information handling program – for example, *Counter for Windows* – to create a graph of the results. Work with the whole class to enter the headings into the chart. Then count the information pictures on the wall and add the number of children next to each heading. Ask the children to come to the computer and type the number in for you. Once all the information has been entered, show the children how it can be changed very quickly into a graph. Talk to them about how the computer can be used to store this and other information. Show them how they can save the information in the program and how they can restore the information without losing it.

Where I live

Joanne and her mum are going to school.

Joanne lives in the middle house.

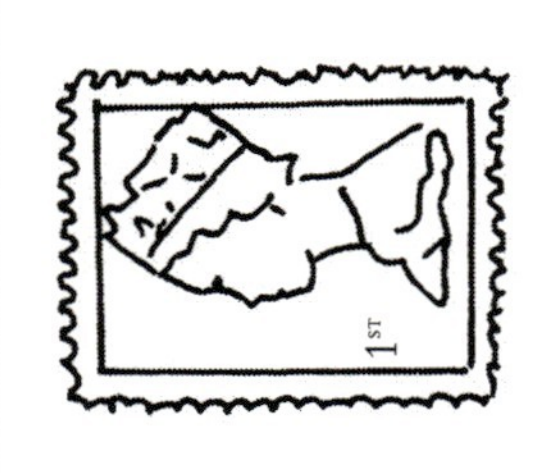

Miss Joanne Smith
6 South Street
Bartley Green
Birmingham
B32 3NT

PHOTOCOPIABLE

Name ______________________________

Where I live

1 ACTIVITY SHEET

Draw a picture of your home.

Draw a picture of yourself on the way to school.

On the back of this sheet, write your name and address.

Name ______________________________

Where I live

Complete the following sentences.

My name is ______________________________

My house/flat number is ______________________________

My street is ______________________________

My town is ______________________________

My postcode is ______________________________

Draw a picture of yourself on your way to school.

On the back of this sheet, draw a picture of your home.

Name ____________________________________

Where I live

Write your name and address on this envelope:

____

____

____

____

I travel to school by (tick the correct box):

Complete these sentences. Use the word bank to help you.

My journey to school takes ______________________________

I live in a ______________________________

WORD BANK

caravan flat house maisonette minutes
mobile home bungalow boat

On the back of this sheet, draw a picture of your home.

Getting to school

CHAPTER 3

TEACHERS' NOTES

This chapter builds on the work of Chapter 1 in terms of developing children's understanding of maps and their ability to draw them and explain what they show. The important points to remember about this chapter are:

- don't try to impose a particular style of map on the children;
- don't expect them to draw a map that looks like a conventional map;
- praise all their attempts because this is an activity in which it is easy to fail and that could have long-term consequences for a child's ability to use and draw maps.

The main activity in the following lesson plan relates to the journey to school. Bear in mind that lots of children may travel to school by car or bus and have little idea of the route taken. By the end of the lesson they will have drawn a map but it is best that they are not told in advance that they will be asked to draw maps as they may worry unnecessarily.

It will be useful to have done some preparatory work. For example, a day or two before the lesson, the children should have looked carefully at the landmarks they pass on their way to and from school. Explain that these landmarks may be houses, shops, road junctions, farms, pubs, churches or other places of worship, post offices, garages or any other features. They should also have tried to remember the order in which they pass these features as they come to school.

Another useful approach is to have asked the children a day or two before the lesson to draw pictures of things they hear on their way to school and things they see on their way to school. These features may include buses, cars and traffic lights as well as people, such as school crossing patrols, police, postal workers and delivery drivers.

Again, they could have been asked to look out for the names of streets that they pass on their way to and from school, which will help in identifying some of the key landmarks around the school. It may also be useful before the lesson to practise the idea of a sequence of places that are passed on a route. So you could ask the children to describe the route from the classroom to the hall to an imaginary pupil who is new to the school. Encourage them to describe the things they would pass on this route, such as other classrooms, the library or the toilets, and even to draw pictures of some of the things passed along the route and then to sequence these. In this way they will begin to grasp the idea of landmarks along a route and how these can help in describing a route to another person.

Getting to school

Geography objectives (Unit 1)
- To draw a simple map of their route to school.
- To understand their sense of place in relation to home and school.
- To describe a route.

Resources

- Generic sheets 1–3 (pages 40–42)
- The pictures from Generic sheet 3 cut out and stuck on card
- Activity sheets 1–3 (pages 43–45)
- Marker pens, paper, crayons, pencils, scissors and glue or sticky tape
- A street map of the local area

Starting points: *whole class*

Tell the children that they are going to listen to a little girl describing her route to school. Show them Generic sheet 1 on an OHP or in an enlarged version and explain that they are all going to follow the route on the map.

Explain that a map shows where places are, such as the school and Joanne's house, and that it shows how to get from place to place.

The text for you to read out is on Generic sheet 2. Bear in mind the following as you read:

- Point out the picture of Joanne's house, where her journey starts.
- Read the first part of her account of her journey until you reach the point where Ali joins Joanne and her mum. Show the children the picture of Ali on Generic sheet 3. Ask a child to come out to stick the picture of Ali on the map in the correct place.
- Read the next sentence of the story. Ask a child to point out the newsagent's shop. Ask another child to come out to draw an arrow on the map to show this next part of Joanne's route.
- Continue the story. Point out the turn to the right and show the picture of the postbox. Ask a child to stick the picture in the correct place on the map. Ask another child to draw an arrow on the map from the newsagent's shop to the pub.
- Continue the story. Point out the picture of the garage on the corner. Ask a child to add arrows to the map.
- Complete the story and show the pictures of the school crossing patrol and the school sign. Ask children to stick them on the map. Ask a child to complete the route map with an arrow.

Now talk the children through the whole route again, using the map and highlighting the key landmarks Joanne passes on her way to school.

Finally, ask children to describe some of the things they pass on their way to school and the names of some of the streets around the school. It is also useful to ask them who they come to school with (a parent, carer or brothers and sisters). This is very important to many young children and may assist in terms of helping them to remember the key landmarks.

Tell the children that they are now going to draw their route to school.

Group activities

Activity sheet 1
This sheet is aimed at children who need more support. They can recognise two or three landmarks that they pass on their way to school, but may struggle with drawing a conventional map. They have to draw pictures of two things they pass on their own route to school.

Activity sheet 2
This sheet is aimed at children who can work independently. They know the names of some of the streets near the school and near their home (for the street in which they live, see Chapter 2). They have to draw pictures of five things they pass on their way to school and draw in arrows linking them (as they did for Joanne's map). Then they have to write on the back of the sheet the names of two streets near the school.

Activity sheet 3
This sheet is aimed at more able children. They understand the idea of a map and are able to draw in and name streets in the correct locations. A street map of the roads around the school would be a very useful aid here.

Plenary session

Share some of the maps of routes to school emphasising the key features of maps:

- They show where places are – for example, street names.
- They show how to get from one place to another – for example, home to school.

Say to the children, 'So now we can see that we have all drawn a map.'

Ideas for support

To help the children grasp these concepts use a series of pictures or photos of places in the school. These should be places such as the classrooms, the hall and the secretary's office. Then talk to the children about the route from their classroom to the hall and get them to sequence the pictures or photos. These are the landmarks they pass along the route.

Help the children's understanding of maps by working on the elements of maps such as direction. Practise ideas of left and right and use simple worksheets showing animals and toy people – from both the front and the back – with instructions to colour the left arms and legs in yellow and the right arms and legs in red.

Other activities to help develop map skills can focus on the use of mazes. Draw several mazes and get the children to draw coloured lines to highlight ways through. Ask them to put a cross at every right turn and a circle at every left turn.

Ideas for extension

Children who are able to draw simple picture maps can progress to drawing simple plans of the classroom, using the correct plan view for objects, and symbols to represent objects. They can also draw a picture of their bedroom and below it draw a plan of the same room showing items such as beds, wardrobe, chairs and a computer in plan view. This activity can be extended to kitchens and/or living rooms, with the emphasis on drawing all items in plan view.

Linked ICT activities

Using the program *My World*, load the file called 'Maketown' from within the program. The program loads with a blank area on the screen and a number of objects, roads and buildings which the children can use to create their own layout of a town or village. The children can resize an object and change the angle before they drag it to an area of the blank screen.

Start by introducing the program to the whole class if possible. A whiteboard is really useful to introduce the program and to allow the children to participate in building the picture. Having already looked at the landmarks and street names in the school environment, start with the school and build some of the main streets onto the screen. The program will also allow you to add the names of the streets to the scene. Give the children the opportunity to add some of the landmarks and objects to build the street scene. Show the children how to save their completed scene and how to print out the final product.

Having completed the scene all together, give the children the opportunity to create their own scene from around the street where they live or a landmark near to them – for example, around the park. Use the printed scenes to create a class display called 'Around our school'.

Joanne's walk to school

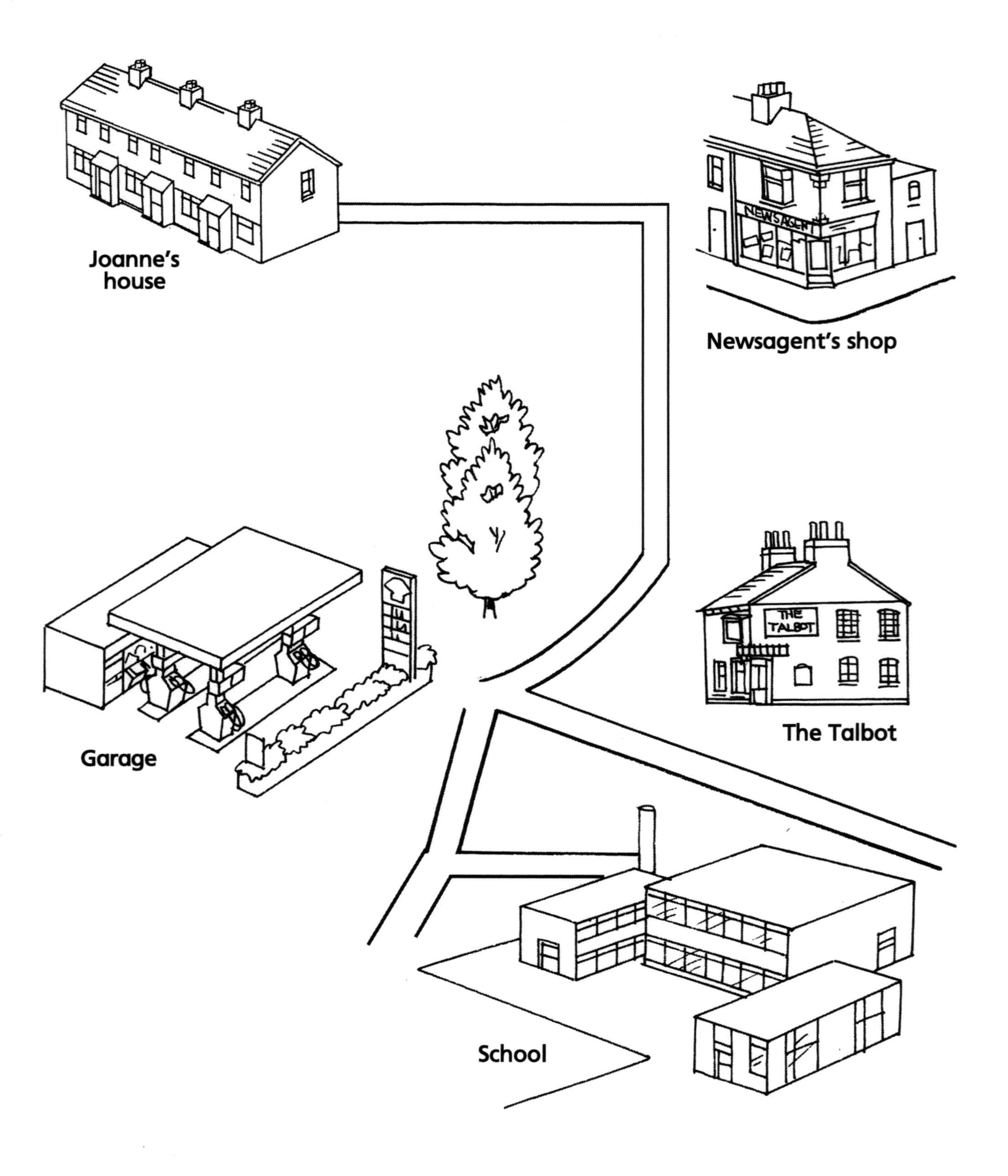

Joanne's walk to school

I walk to school each morning with my mum.

We leave home and walk a little way along the street until we meet Ali, who walks with us.

All three of us then turn right and walk along the road past the newsagent's shop.

Then we walk past the postbox.

We see a tall tree on the right.

Then we pass a pub called The Talbot.

Next, we come to the garage on the corner.

After the garage, the crossing warden helps us to cross a busy road so that we can reach the road with a school sign in it.

Finally, Mum leaves us when we go into school.

Joanne's walk to school

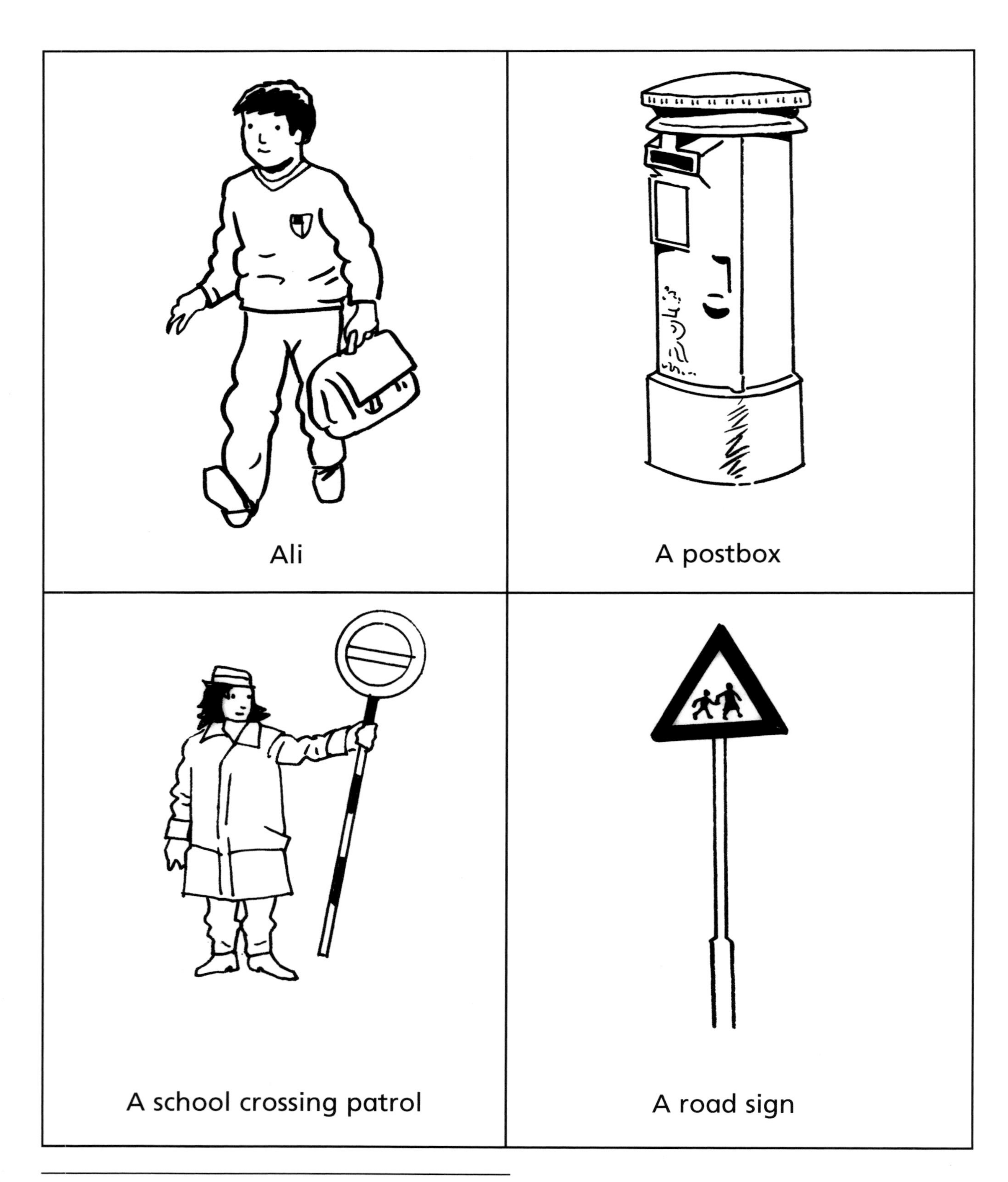

Ali

A postbox

A school crossing patrol

A road sign

Name ______________________________

Getting to school

Draw pictures of two things you pass on your way to school.

Home

I see this on my way to school

I see this on my way to school

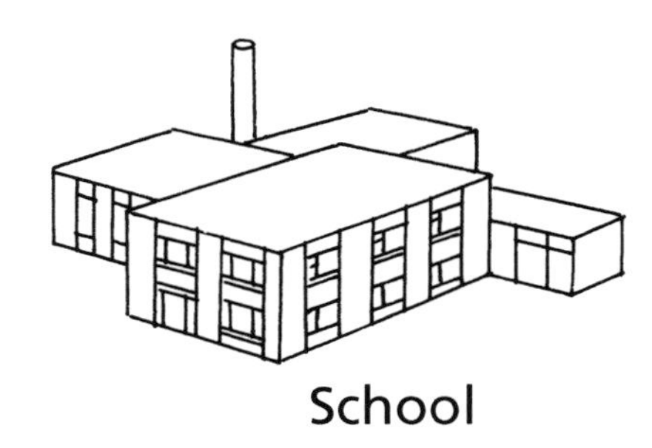
School

Name ________________________________

Getting to school

Draw pictures of five things that you pass on your way to school.
Then, draw arrows to connect these things.
On the back of this sheet, write the names of two streets near to school.

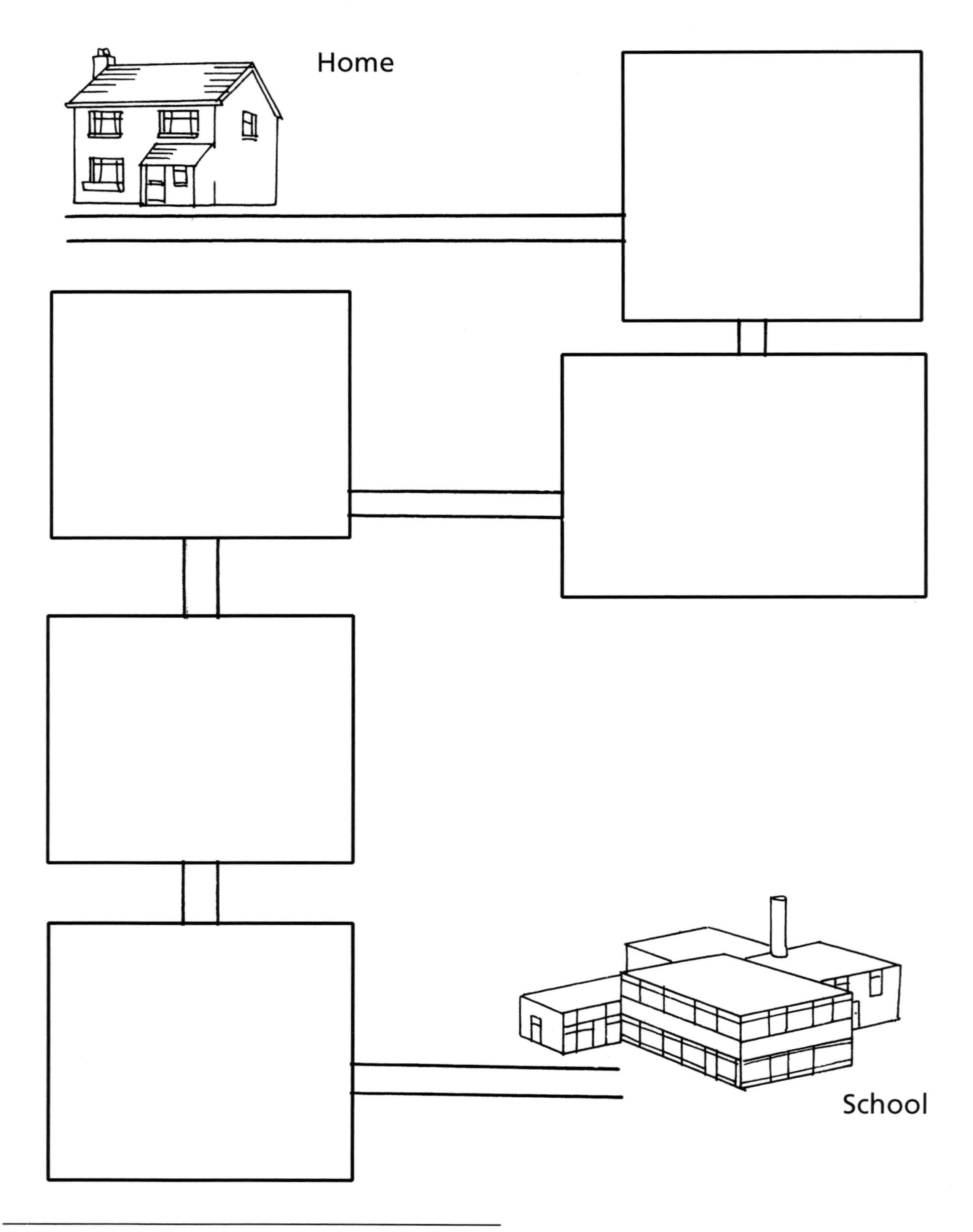

Name ____________________________________

Getting to school

Complete this picture map to show streets and houses, shops and other things that you pass on your way from home to school.

Home

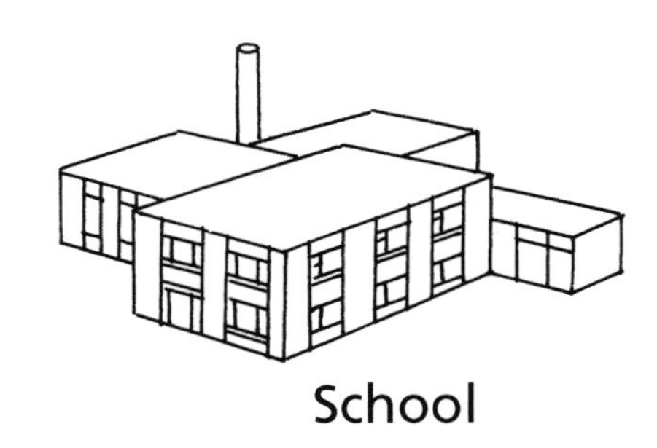

School

Around our school

TEACHERS' NOTES

The aim of this chapter is to encourage children to look more closely at their local environment and to begin to think about features that are physical (natural) and things that are human (made by people).

Knowledge of the local area

Ideally, the children should have experienced a class walk around the local area. They should understand the difference between physical (natural) and human (made by people) features. It is useful to take photographs of the features you see to use back in the classroom for sorting, discussion and display purposes. It is even better if you can use a digital camera because the results can be displayed on the computer as soon as you return to the classroom. It is also possible to use these images to get close-up views of some of the features.

Guidelines for walks around the local area

- Consult the school and local authority guidelines that deal with taking children out of school. These will cover all the key points of administrative detail and also ensure that you have completed successfully all the necessary risk assessments.
- Check that you have enough adult helpers – again, guidelines will detail how many are essential. Aim for at least one adult for every ten children, although more is better.
- Walk the route yourself first, and highlight potential danger points, such as road crossing points, and what you intend to do about them.
- Brief all your adult helpers as to what you want the children to do. Some helpers are content to look after a few children but may not be clear on the purpose of the activity and exactly what you want each child to do and to have achieved/experienced by the end of the visit. Ensure they understand why they are going out and what you expect of them.
- Make sure that you gain parental approval if this is deemed necessary by the headteacher, school and local education authority.
- Take adequate first-aid equipment with you and make sure that you have enough adults who know how to use it.
- Make sure that the children understand exactly why they are going out and what you want them to do, as well as how you want them to behave.
- Ensure that you are clear about what you will do if the weather is poor on the day or becomes poor during the outing.
- Don't aim to be out for longer than about 45 minutes.

Physical features

The physical (natural) features surrounding your school will, of course, vary according to the location of the school. See page 47 for a list of some features you might encounter.

Human features

The human features in your area include all the things that have been constructed by people. These will include the roads, traffic lights, bridges and buildings as well as other features that might seem, at first sight, to be physical features, such as parks and farmland. See page 47 for a list of some features you might encounter.

It is important that children begin to recognise the different types of buildings in their local area and begin to understand the different purpose of each. Building types include houses (terraced, detached, semi-detached, high-rise apartments, flats, bungalows, cottages), shops, garages, pubs, hotels, churches, mosques, temples, supermarkets, cinemas, theatres, restaurants, cafes, police stations, fire stations, ambulance stations, offices, town halls, village halls, sports centres, museums, libraries, canals and railway stations.

Physical features

Bay A curved part of the seashore.

Beach An area of sand or pebbles at the seashore.

Cliff A steep, high rock face, especially along the seashore.

Estuary The wide mouth of a river where it meets the sea.

Flood plain The flat area alongside a river over which it floods.

Forest A large area of trees.

Headland An area of higher land that sticks out into the sea along the coast.

Hill An area of higher land, but smaller and lower than a mountain.

Lake or **Loch** A large area of water. Some are completely surrounded by land but others are open to the sea at one end.

Moor An area of grass, heather or bracken often used for the grazing of animals.

Mountain An area of high, steep land. Some of the highest mountains are covered in permanent snow and ice.

Pond A pool of water. This can be made by people – for example, farmers create ponds for their animals.

River An area of fresh water that flows from high land to low land. Many rivers flow to the sea.

Stream A small river.

Tributary A stream or river that flows into a larger one.

Valley Low-lying land between hills.

Human features

Canal A water channel built across land to join two areas of water.

City A very large urban area.

Cold stores Refrigerated buildings used to store perishable foods such as meat.

Dam A strong wall built to hold back a river.

Dock A place on a river, canal or coast where ships can stop to load and unload.

Electricity pylon A large metal structure built to carry electricity cables across the area.

Environment The surroundings of people, plants and animals.

Ferry A boat for carrying people and goods across a river or sea.

Hamlet A small group of houses with very few services such as shops.

Housing estate A planned area where houses are built and services provided.

Industrial estate An area of land occupied only by factories and smaller industrial premises.

Lock Part of a river or canal, with gates at each end. The level of water can be adjusted to allow boats to move up or down the river or canal.

Market town A town with an old marketplace, usually in a central square.

New town Planned settlement; many built in the 1960s and 1970s.

Port Town on the coast where goods and people can enter or leave the country by ship.

Refinery A place where crude oil is made pure. Petrol, diesel oil and other products are made from crude oil.

Reservoir A lake that builds up behind a dam. It is used for collecting and storing water.

Services Things people need, such as shops, schools and hospitals.

Settlement Place where people live.

Skyscraper Very tall building with many storeys.

Terraced house House built in a row, joined to each other.

Town Settlement larger than a village, but smaller than a city.

Village Settlement in the countryside. Many villages have services such as shops, a church, a pub, a post office and, perhaps, a doctor's surgery.

Around our school

Geography objectives (Unit 1)
- To recognise some of the physical and human features in their locality.
- To understand some of the ways in which features are used.

Resources

- Generic sheets 1 and 2 (pages 50 and 51)
- Activity sheets 1–3 (pages 52–54)
- Photographs of human and physical landmarks in your local area; alternatively, pictures of similar features
- A4 paper
- Clipboards

Starting points: *whole class*

Take the children on a short walk in the area around the school (see page 46). As you walk, pause at key landmarks such as postboxes, shops, road junctions, trees or traffic signs and ask the children if these are physical or human features. They could record these on a map or draw some of the more important ones.

Tell the children that they are going to look at pictures of things they pass on their way to school each day. One at a time, show them the pictures of human and physical landmarks, including those on Generic sheets 1 and 2 on an OHP or in enlarged versions.

Ask questions such as:

- What is this?
- What do people use it for?
- Is it close to or far away from school?
- What type of building is this?
- Do you pass one of these on your way to school?

Look at all the pictures again. Tell the children that you are going to sort them in some way. Sort them into two groups – human and physical features. Ask the children to guess how they have been sorted. Share ideas on how things in the groups are different.

You could then sort the buildings into different groups, such as shops and houses. Invite the children to tell you any other ways the pictures could be sorted. Next, ask them to tell you which things are near to the school. Put these in one group. Which things are further away? Which are furthest away? Invite them to place the pictures in order from nearest to the school to furthest away. You could do this by attaching the pictures in the correct order on the board.

Tell the children that they are going to think about the things they pass on the way to school. Remind them of the items in the pictures on the two generic sheets. If necessary, give them copies of the sheets. Say that you want them to think about which thing they see first when they leave home and which thing they see last when they reach school.

Group activities

Activity sheet 1
This sheet is aimed at children who need more support. They have to tick different physical and human features that they pass on the way to school. Then they have to draw pictures of two things they see on their way to school. They are asked to cut out the pictures and put them in the order that they see them.

Activity sheet 2
This sheet is aimed at children who can work independently. They have to identify the different physical and human features they see on the way to school. On a separate sheet of paper, they have to draw pictures of three things they see on the way to school and then put them in order.

Activity sheet 3
This sheet is for more able children. They have to complete a chart showing what they see and roughly the quantity they see. They then have to draw and label five things that they see on the way to school, in the order in which they see them.

Plenary session

Share the responses to the activity sheets. Talk about whether there are more physical features that the children see than human features, or vice versa. Why is this the case? Where would the children have to go to see more human/physical features – for example, the seaside?

Ideas for support

Before they start the activity sheet, ask the children to draw *people* they see on the way to school, because they will remember people better than things. Once they have drawn the people and discussed them, ask them to think about **where** they saw them (such as near the chip shop).

Produce a display of photographs of the area around the school. These could be photographs taken by you or by the children on a walk around the area. Use the display to distinguish between physical and human features. If possible, record a short piece of video of your own journey to school or that of a colleague, and show this to the children. Use it to show the children the different features passed and, again, reinforce the distinction between physical and human features by pausing on the key features for a few moments.

Ideas for extension

Ask the children to sort the pictures from Generic sheets 1 and 2 into three sets:

- things found only in country areas;
- things found only in towns;
- things found in both town and country.

Ask the children to imagine that there is a new member of the class who needs simple instructions on how to get from the school to the nearest shop or other main feature. Ask the children to draw a simple 'picture map' of all the physical and human features that this child would pass from leaving the school gate to arriving at the shop/other main feature.

Linked ICT activities

Create a wall display of photographic images of different physical and human features and talk to the children about the different words and sentences that they have used to describe them when talking about them. Using a simple word processing program such as *Talking Write Away* or *Textease*, work with the children to create words and sentences to label the display. Encourage them to use different font styles, colours and sizes to create the labels for landmarks and signs.

Give the children the opportunity, where possible, to use a digital camera to take images of what they see from their level.

Around our school

House

Pub

Mini-mart

Newsagent's

Fish and chip shop

Block of flats

Cottage

Office blocks

Police station

Farm

Church

Mosque

Around our school

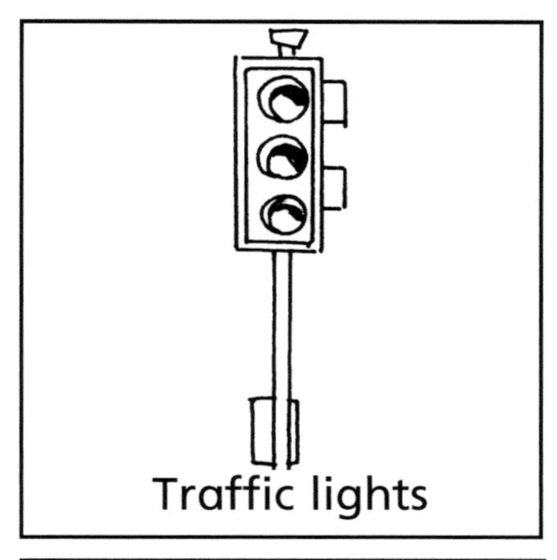
Traffic lights

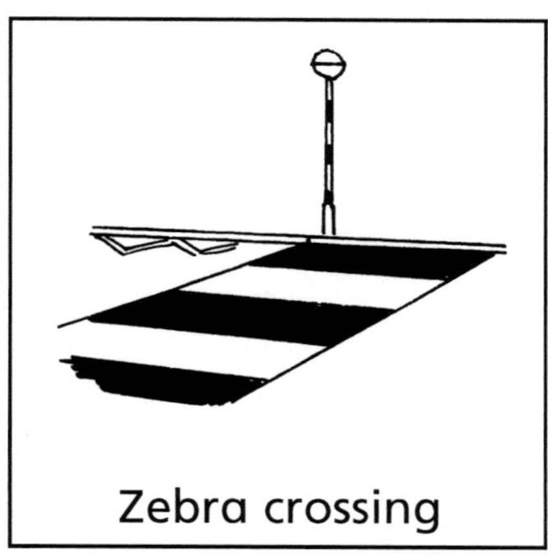
Zebra crossing

Lamppost

Phone box

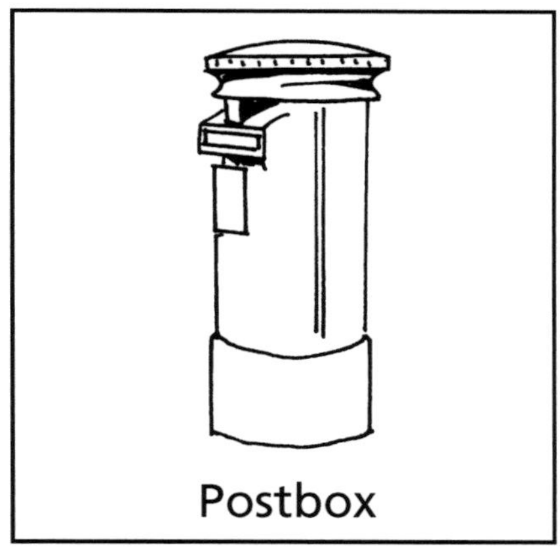
Postbox

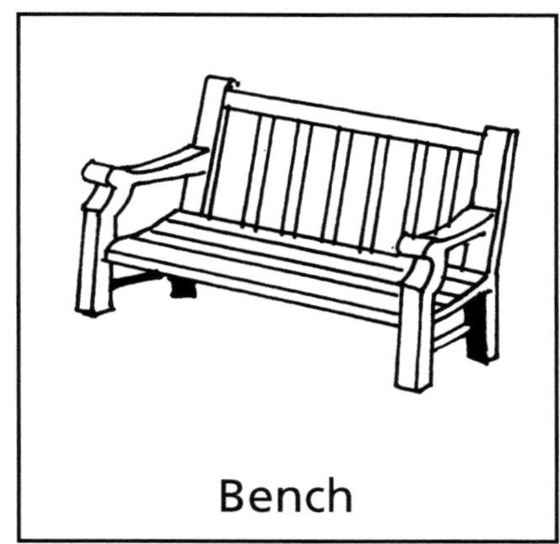
Bench

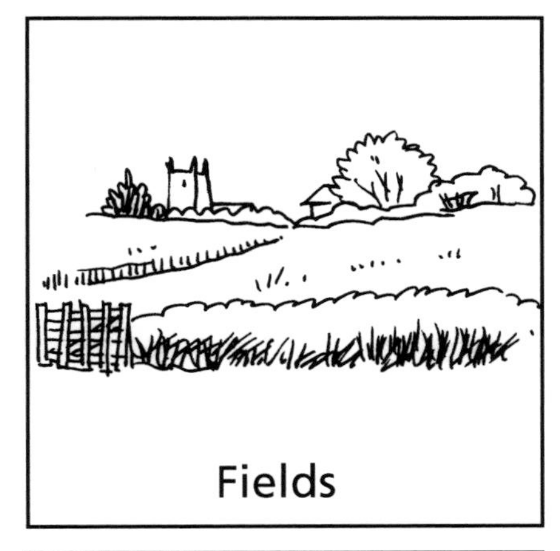
Fields

Hills

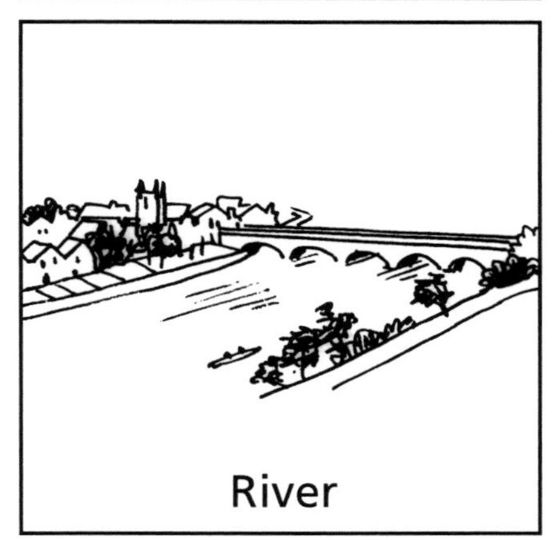
River

Woods

Mountains

Park

Name ______________________________

Around our school

Put a tick by the things that you see on the way to school.

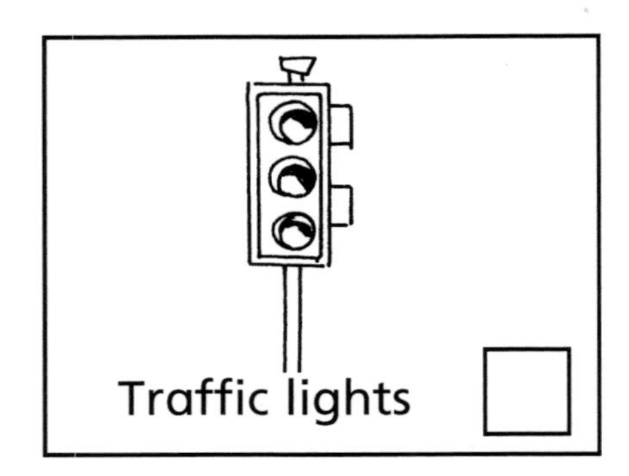

Traffic lights ☐

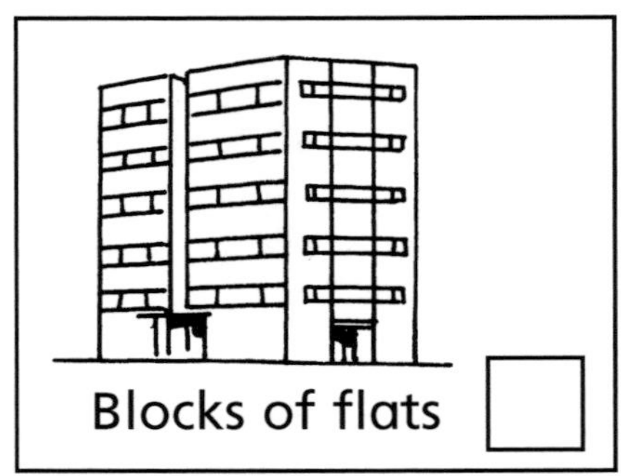

Blocks of flats ☐

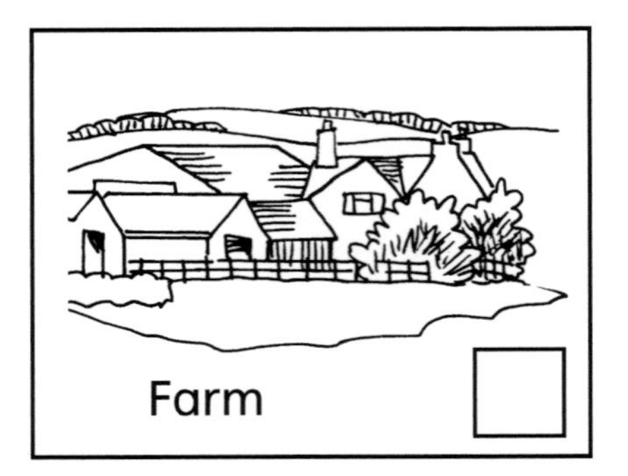

Farm ☐

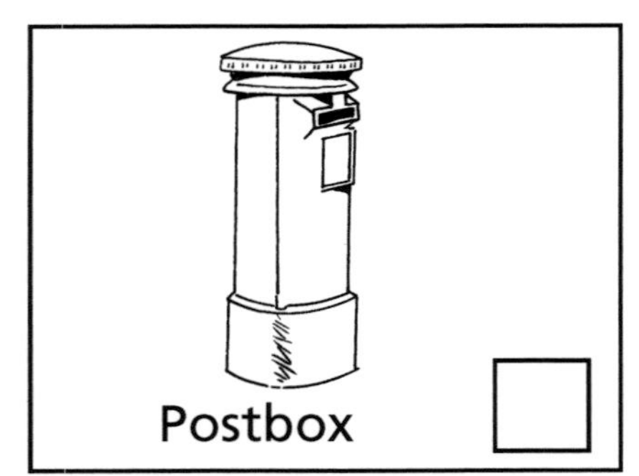

Postbox ☐

Chip shop ☐

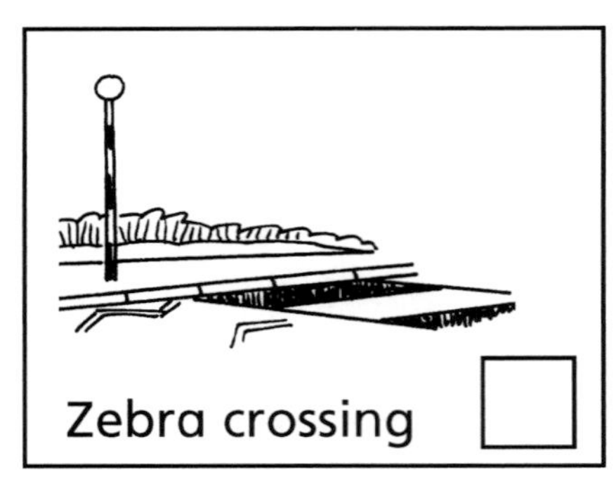

Zebra crossing ☐

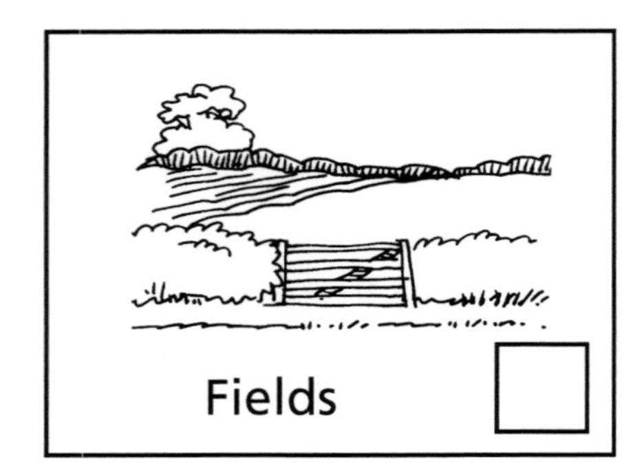

Fields ☐

Police station ☐

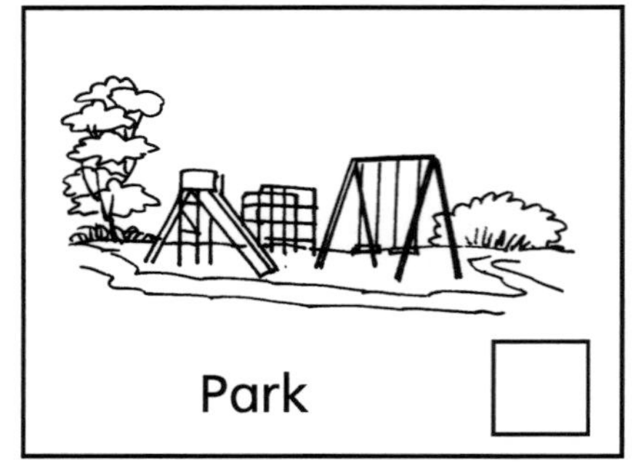

Park ☐

Draw two things that you see on the way to school.

Cut out your pictures and put them in the order that you see them on the way to school.

Name ______________________________

Around our school

Put a tick by the things that you see on the way to school. ✓

- police station ☐
- house ☐
- office block ☐
- chip shop ☐
- block of flats ☐
- grocer's ☐
- newsagent's ☐
- cottage ☐
- fire station ☐
- pub ☐
- church ☐
- lamppost ☐
- hill ☐
- zebra crossing ☐
- river ☐
- bench ☐
- park ☐
- postbox ☐
- mountain ☐
- phone box ☐
- field ☐
- traffic lights ☐

On another sheet of paper, draw three things that you see on the way to school. Cut out your pictures and put them in the order that you see them on the way to school.

Name ______________________________

Around our school

Complete the chart to show what you see on the way to school and how many of each you see.

What I see

What I see	Only 1	Between 2 and 5	Between 6 and 10	11–20	More than 20
house					
lamppost					
fish and chip shop					
farm					
shop					
postbox					
police station					
river					
library					
block of flats					
stream					
telephone box					
cottage					
field					
fire station					
pond					
post office					

How many I see

On the back of this sheet, draw and label five things that you see on the way to school, in the order you see them.

Our surroundings

TEACHERS' NOTES

Many people walk through their surrounding area as if they have their eyes closed. Children do hold views about the nature of their surroundings, but often they don't express them, or they simply reiterate something they have heard from parents, carers or peers. The challenge is to get them to think about their environment and not simply to accept that it is something over which they have no control.

Where to start

There is often much that children can do to shape the nature of their surroundings. Some examples are listed below:

- They could hold regular meetings of a litter 'task force' to clean up the school and its grounds, with the aim of keeping them litter free.
- They could produce posters and signs encouraging other children not to drop litter. They could then walk around the school and decide on the best locations for the posters (having discussed size, colour, lettering, image and layout).
- They could establish their own area within the school grounds where they might be able to plant some bulbs or bushes and create a garden. This could be with the aim of turning a boring, 'nasty' area of concrete or tarmac into a nice place. On a larger scale, this could expand to the creation of a shallow pond and water feature with a suitable fence. Encourage the children to plant bushes with berries for the birds. Help them to set up bird-feeding tables and to plant bushes such as buddleia, which will attract wildlife and insects. (Check with the school's safety policy with regard to ponds and plants.)
- They could think about the different ways in which they could improve the playground. What additional features would they like to see? Then ask them to explain why they selected their particular features.

Other ideas for exploring

- Start with the classroom. Ask the children to say which areas they think are nice places and which are nasty. Ask them to say why they feel these places are nice or nasty (nasty might be near a litter bin; nice might be near the radiator or the window). Ask the children to suggest how nasty places (such as the sink) might be improved (by people taking more care in the sink or when putting litter in the bin).
- Take a series of photographs of nice and nasty places around the school and the local area. Make sure that you keep a note of where you took the photos (street, house number, side of road).
- Produce a series of photos that show close-ups of particular parts of houses, shops or street signs. Each photo may make no sense on its own, but the task for the children is to spot the places (say ten) as they walk around the area with you.
- Produce a series of photos that show patterns in the environment. These could be patterns of slabs in a footpath, patterns of bricks in a wall, patterns of windows in a frame or patterns of lights along a road. Again, as you walk around the area with the children, challenge them to spot these patterns, together with new patterns that they can then photograph.

Our surroundings

Geography objectives (Unit 1)
- To describe the features of the local environment.
- To express views on the features.
- To know that changes occur in the locality.

Resources

- Copies of Generic sheet 1 (page 58) one per group of two or four children
- Six copies of Generic sheet 2 (page 59) with the circles cut out and coloured green for 'nice' and red for 'nasty'
- Activity sheets 1–3 (pages 60–62)
- Double-sided sticky tape
- Camera

Starting points: *whole class*

Tell the children that they are going to look at places around them and decide if they are nice places or nasty places.

Encourage them to look around the classroom and to suggest which are nice places (display boards) and which are nasty places (the area near the rubbish bin). Discuss what makes a place nice or nasty. Ask questions such as:

- Does a place have to be clean to be nice?
- Is the amount of litter important?
- Is the tidiness of the place important?
- Does the colour or shape of the place affect whether it is nice or nasty?
- Can a cold place be nice or does it need to be warm?

Now show the children the pictures from Generic sheet 1. Ask them to describe what they can see in each picture. Tell them that you are going to sort the pictures into groups. Sort them into places you like and places you dislike. Can the children guess the basis for your sorting? Ask them if there are places in your 'nasty' group that they think are nice. Discuss the idea of different people liking and disliking different things. Ask questions such as:

- Why do some people like gardens and others dislike them?
- Why do some people like noisy places?

Now show them the circles from Generic sheet 2 and explain that they are going to walk around the school and decide if places should get a nice sticker or a nasty one. Explain the need to be quiet when doing this.

Walk around the inside of the school, stopping at key places and deciding with the children if a sticker is appropriate at that place. Nice places might include the hall, the entrance area and the library. Nasty places might include the cloakroom area (especially if coats are on the floor), places where litter can be found and places with steep steps or other potential dangers. Let the children stick the circles in all these places, and take photographs of each one.

Next take the children outside the school building into the playground. Again, discuss which are nice places (flowerbeds and play areas) and which are nasty places (the school rubbish bins and playground areas with litter). Again, let the children stick on the circles and take photographs.

Tell the children that they are now going to record nice and nasty places that they know.

Group activities

Activity sheet 1
This sheet is aimed at children who need more support. They can recognise nice places that they see on a regular basis. They have to draw (and name) a nice place and a nasty place that they pass on their way to school. They have to write as many labels as they can.

Activity sheet 2
This sheet is aimed at children who can work independently. They can identify nice and nasty places that they see or visit. A copy of Generic sheet 1 would help them. They have to draw one nice place and one nasty place. They have to write why they think it is a nice/nasty place.

Activity sheet 3
This sheet is for more able children. They have to complete a chart showing how many of the different nice and nasty places listed they see or visit. They have to draw pictures of two nice places and two nasty places.

Plenary session

Display the photographs the children took around the school and add two or three that you have taken in the local area. Discuss how nice or nasty these 'new' places are. Discuss what makes them nice or nasty. Ask questions such as 'How could we make this nasty place nice again?' (For example, by collecting litter from the playground or playing field.)

Ideas for support

For children who have difficulty deciding what is nice and what is nasty, put up a wall display of typical pictures to help them.

You could also put up a wall display of a section of road and pavement with simple shop fronts and a few people. Help the children to cut out and colour simple shapes of cars, lorries, buses, vans and motorcycles and then ask them to stick the shapes in the correct place on the wall display. Encourage them to talk about why this might be a dangerous place (speed of traffic, problems in crossing the road, car exhaust fumes). Simple 3-D models of the vehicles could be used instead of the cut-out shapes to give the display a more realistic feel.

Ask the children to draw three pictures that show a person walking around three places either in the school or near the school. Two of the places should be nice and one should be nasty. Ask them to sequence the pictures on the basis of how close each place is to the school.

Ideas for extension

Ask the children to locate photographs taken by themselves or by you on a large-scale street plan of the area around the school. Tell them to stick the photographs in place and add a short description of why it is a nice or nasty place.

Ask them to devise a route around their local area, starting from the school, on which they could take a new pupil in the class to introduce them to the nicest parts of the area. The route should include descriptions of which ways to turn (left, right, or north, south) and the names of the streets that the children would walk along.

Study some of the photographs of nasty places and then discuss with the children how the area could be improved – for example, litter bins and litter patrols could be provided and signs and posters could be put up to encourage people not to drop litter. Ask them to write an action plan for improving one of the areas.

Linked ICT activities

Tell the children that they are going to create their own sign for one of the nasty places they have identified, using a graphics program such as *Dazzle*. The sign should try to help keep the place nice – for example, 'Please throw your litter in the bin' or 'Please keep your dog on a lead'. Talk to the children about the kinds of signs that we already see in our environment that are designed to keep the environment clean and tidy.

Talk to the children about what their sign will say before they start to use the computer. Show them how to use the program's text tool to write their sign. Show them how to change the size of the font and the type and colour of the font. Let them write their sign and, using other paintbrush tools, show them how to illuminate and add colour to the background and letters to create a finished sign. Print out the signs and display them in the classroom.

Our surroundings

Our surroundings

Name ______________________________

Our surroundings

1 ACTIVITY SHEET

Draw one nice place and one nasty place that you pass on your way to school. Write their names in the boxes.

My nice place is ______________________________

My nasty place is ______________________________

Write some labels for your pictures.

Name ______________________________

Our surroundings

Put a tick by those nice and nasty places you see or visit often. ✓

Nice places		Nasty places	
Play area	☐	Place with lots of litter	☐
Park	☐	Busy main road	☐
Trees	☐	Dangerous canal	☐
Fast food shop	☐	Noisy place	☐
Games shop	☐	Crowded pavement	☐
Church/mosque	☐	Other nasty place	☐
Other nice place	☐		

Draw one nice place and one nasty place you see often.

Nice	Nasty

On the back of this sheet, write why you think your nice place is nice and your nasty place is nasty.

Name ____________________

Our surroundings

Complete the chart to show how many nice and nasty places you see or visit. Use green for nice and red for nasty.

What I see

What I see	Only 1	2–5	6–10	11–20	Over 20
play area					
park					
trees					
fast food shop					
games shop					
church/mosque					
other nice place					
rubbish dump					
busy road					
dirty canal					
noisy place					
sewers					
other nasty place					

How many I see

On the back of this sheet, draw two nice places and two nasty places. Write what you like and dislike about them.

Jobs in the local area

TEACHERS' NOTES

Job classification

Most modern jobs can be classified as:

- **primary** – includes mining, agriculture, fishing and forestry;
- **secondary** – includes any type of manufacturing industry, from cars to computers, and includes construction;
- **tertiary** – service industries, which can be further subdivided into:
  - **quaternary** – jobs based on transport, from bus drivers to train staff to taxi drivers;
  - **quinary** – service jobs in education, the health service and social services.

Today most people work in some form of service industry. The service industries can also be sub-divided into:

- **producer services** that exist to serve other organisations; these include banks, insurance companies and other financial services;
- **consumer services** that are provided by retailers (who provide services for shoppers), hotel chains and leisure organisations;
- **public services**, which include the health service, fire, police, education and social services.

One way to help children understand about different jobs is to look at the production of a book. It involves:

- growing the trees, which are felled before being sent to the factory (this tree cultivation is a primary industry);
- processing the wood from the trees into wood pulp and then into paper at a factory (this is a secondary industry);
- publishing and distributing the book – the author types the manuscript into a word processor, editors prepare the book, artists draw maps and diagrams, printers produce the final pages (a secondary, manufacturing function), lorries take the books to warehouses, sales staff sell them.

The changing nature of jobs

It is important to realise that local jobs can change quite radically over a short (20-year) period of time. The two pie charts on page 64 show the percentage of the UK's employed population in primary, secondary and tertiary industries in 1971 and 2002. The charts show clearly how the percentage of the UK population who work in factories in the manufacturing industry has declined dramatically over 31 years, from 33 per cent to 19 per cent. This decline is sometimes called **deindustrialisation** and it is still continuing in some parts of the UK.

This decline is due to:

- increased competition from other countries, usually due to the lower wages paid there;
- out-of-date factories and machinery in some parts of the UK;
- a world surplus of some products, such as steel;
- increased automation – better machinery needing fewer people.

The effects of this deindustrialisation are felt in places in the UK where many factories that make steel, ships, cars, bicycles, motorcycles, clothing, electrical goods and chemicals have been closed down. Many people have been made redundant and have found it hard to secure alternative employment. Even when new jobs are available, they are often temporary, part-time, poorly paid or involve long hours. They may require retraining in a completely new industry.

So the children need to know:

- what jobs are available locally;
- what jobs used to be found locally but are no longer available;
- that people now need to change their jobs several times during their working life because things, especially technology, are changing so rapidly.

You need to be sensitive to the fact that some children's parents or carers may be made redundant and be unemployed for a certain period of time.

However, not all industries are declining. There are some new growth industries in the UK. These are usually **hi-tech industries,** meaning that they use

The changing nature of jobs in the UK

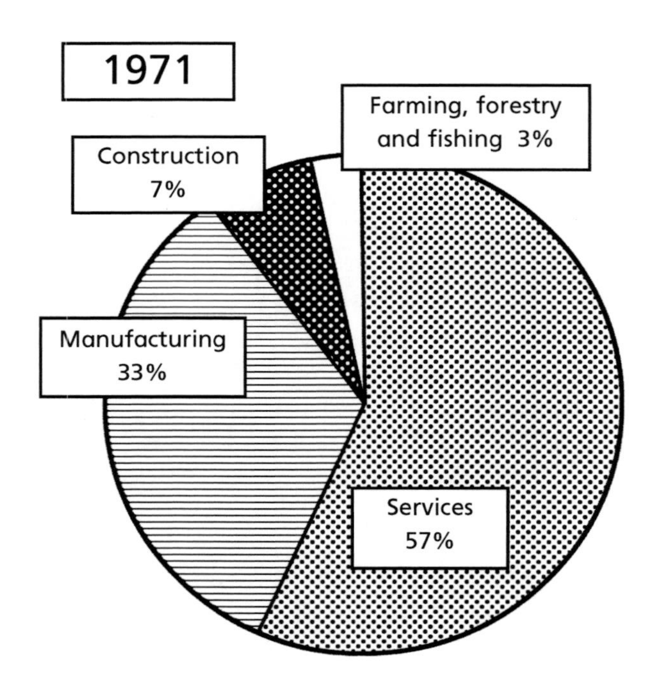

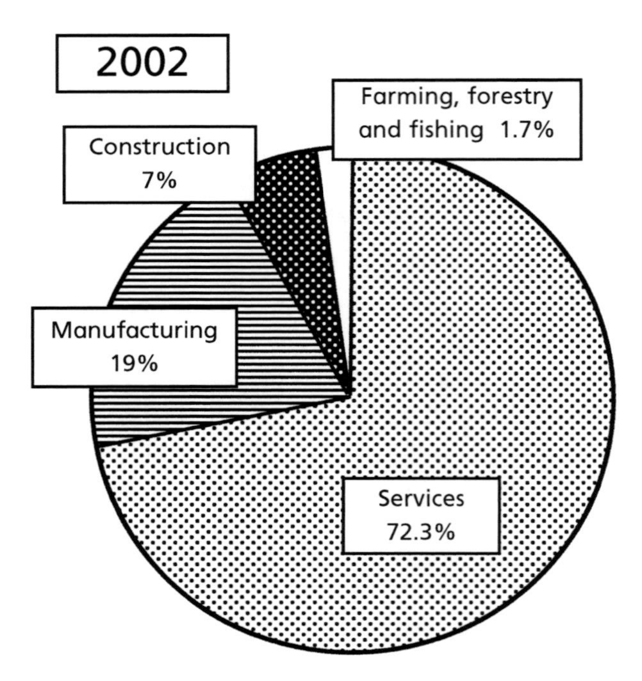

the most modern technology to produce goods such as computers or mobile phones.

Employment in rural areas

A decline in the number of jobs available is also a feature of many rural areas of the UK. In part, this is due to a long-term decline in farming, which is a very automated industry using a lot of machinery but employing fewer and fewer people. Also farms are getting bigger in order to survive so there are fewer farmers. In some places, people have left the countryside and have gone to find work in towns.

Another reason for the decline in the jobs available in the UK countryside is the closure of:

- cottage hospitals and clinics (to concentrate on large hospitals in towns);
- small village schools;
- libraries in villages and small towns;
- bus services in some areas where it is expensive to provide a service to remote areas where relatively few people live.

The exceptions to this decline in the countryside are in those areas where there is **counter-urbanisation**. This is where people leave the towns and cities and choose to live in villages in the countryside. Sometimes this counter-urbanisation is due to the movement of retired people, sometimes it is due to people who can work from home and so choose to live in the quiet, cleaner environment of the countryside. As these people move into the countryside, some new jobs may be created – for example, in building new houses or converting old ones, or in providing services such as shops, pubs or restaurants.

Jobs in the local area

Geography objectives (Unit 1)
- To identify some of the uses of land and buildings in their locality.
- To understand that these uses are linked to the work people do.

Resources

- Generic sheets 1 and 2 (pages 67 and 68)
- Activity sheets 1–3 (pages 69–71)
- Sticky tape or glue

Starting points: *whole class*

Tell the children that they are going to look at pictures of people doing different jobs. They should think about whether there are people in their local area who do jobs like these.

Show one of the pictures from Generic sheet 1 on an OHP or in an enlarged version. Ask questions such as:

- What is this person doing?
- Where do they work?
- Is there a place near you where people do this work?

Ask similar questions about two other pictures of jobs that can be found locally, such as working in a bank and driving a bus.

Now show a picture of a job that is not found locally. In an urban area, this would probably be one of the farming pictures (feeding cattle, driving a tractor). In a rural area, the jobs not found locally might be the manufacturing ones (sewing in a factory, making sweets). When the children understand that these jobs are not found locally, ask questions such as 'Where in the UK would we go to find people doing these jobs?'

Next show Generic sheet 2 and explain that the jobs that people do can be divided into three main groups – people who help us, people who make things and people who grow, catch or cut things. Talk about the examples on the sheet.

Tell the children that they are going to put the pictures from Generic sheet 1 into the correct columns on Generic sheet 2. Go through each of the pictures and ask them in which column it belongs. Let some of them come out and stick the picture in the column. Discuss each one and ask for the reasons behind the classification.

If possible, a visit to a local supermarket is really worth while. Encourage the children to talk to the different people about their work, what they like about it, what hours they work, what clothing they have to wear and what different things they do. Let the children take photographs of the people they interview (with their permission) and possibly record the interviews on tape for use back in the classroom. Remember that the guidance about taking children on an outing in Chapter 4 (page 46) still applies. In addition, you need to obtain permission from the supermarket and explain the purpose of your visit, how many will be in the group, the ages of the children, how many adult helpers you will have, what time you would like to arrive and how long you would like to stay.

Tell the children that they are now going to think about the jobs that people do in their local area and where these jobs are done.

Group activities

Activity sheet 1
This sheet is aimed at children who need more support. They are able to identify some key jobs in the local area. They have to choose the correct names from the word bank and write them under the picture. They also have to write where the people work.

Activity sheet 2
This sheet is aimed at children who can work independently. They have quite a good idea of the jobs that people do in the local area. They have to know the correct names for the jobs in the picture and write them under the picture. On the back of the sheet, they have to write a sentence about where the people work.

Activity sheet 3

This sheet is aimed at more able children. They have to identify the activities taking place, and then add to their lists on the basis of the earlier discussion. Copies of Generic sheet 1 would be useful here. They also have to think about and list some jobs in a non-local area.

Plenary session

Share some of the results of the work of each group with the rest of the class and recap the types of jobs to be found locally. Ask:

- Where do these jobs take place locally?
- Where are the shops?
- Where is the library?
- Where is the nearest other school?
- Where is the police station?
- Are there jobs that local people do that we have not included?
- In a village, why is the shop in the centre? (Easy to find; attracts maximum passing trade.)
- In a suburb, why do we find shops near the houses? (Convenient – saves travelling to distant shops.)

Ideas for extension

Encourage the children to think about the types of equipment different people need in order to do their jobs. A visit by the local fire brigade or Fire Safety Officer would provide specific examples of items such as helmets, breathing apparatus, fireproof clothing, axes and torches. Encourage the children to describe how and when each item of equipment would be used. The same types of activity can be undertaken with a visitor from the police or the health service.

Encourage the children to think about local people who do not have a job. What do they do? How are they helped to find a job? How are they helped while they are looking for a job? Where do people go to look for a job?

Ask the children to think about people who work but are not paid for their work, such as voluntary charity workers, volunteer workers in schools and parents or carers in the home. Ask them to write about all the jobs that these people have to do in the course of a day. How could the children help to reduce some of these jobs?

Ideas for support

Discuss the pictures on the activity sheets with the children. Point out the policeman, the firefighter and the teacher. Talk about what they are wearing and the job they are doing.

Encourage the children to play with a shop in the classroom, 'buying' goods, receiving change and taking away the goods. (They could use simple cakes, tarts, or fruit made from salt dough – flour, water and salt work well.) In addition to helping the children to think about number and the change they will get, encourage them to think about the people they find in the shops. Talk about the different jobs that people in shops do, such as taking the money at the checkout, putting the goods on the shelves, cleaning the floors, baking the bread or cutting the meat and collecting the trolleys.

Linked ICT activities

Having discussed with the children the different types of jobs that people do, talk to them about the different types of jobs that they do around the school and classroom – for example, collecting litter and tidying books away. Work with them to create a 'We need you' poster. Using a simple word processing program (for example, *Talking Write Away* or *Textease*), ask them to think of a job around the school/classroom that they may need someone to do and to write a list of things which the person doing that job would need to be able to do. For example, 'We need you to put the PE equipment away. You must be strong. You need to lift and carry.' Print out the completed text and add it to either a picture that the child has drawn of someone doing the job, or a digital picture that they have taken of another child doing the job.

Jobs in the local area

Jobs in the local area

People who help us	People who make things	People who grow, catch or cut things
A teacher	A car assembly line	Planting a tree

Name ____________________

Jobs in the local area

Look at the picture.
Using the word bank, list all the jobs that you can see.

The jobs I can see are:

WORD BANK

police officer   firefighter   teacher

Write where these people work:

PHOTOCOPIABLE

Name ______________________________

Jobs in the local area

Look at the picture.
List all the jobs that you can see in the picture.

The jobs I can see are:

____

____

____

____

____

On the back of this sheet, write a sentence about where these people work.

Name ___________________________

Jobs in the local area

Look at the picture. On the back of this sheet, list all the jobs that people are doing.

Then add to the list all the jobs that you can think of that people do in your area.

Then list the types of jobs that would not be found in your area.

Leisure time

TEACHERS' NOTES

What is leisure?

'Leisure time' sounds as if it is a simple idea to investigate, but within geography leisure has become intertwined with recreation, sport and tourism. So it is as well to be clear from the start about what we mean by leisure. Leisure is:

- time free from work and other obligations;
- things we choose to do in our spare time.

Types of leisure

- **Passive leisure** means sitting around relaxing.
- **Recreation** includes some form of activity.
- **Sport** is an activity with a formal set of rules.
- **Tourism** involves travel away from the home environment.

The relationship between passive leisure, sport, recreation and tourism is best shown by the diagram below.

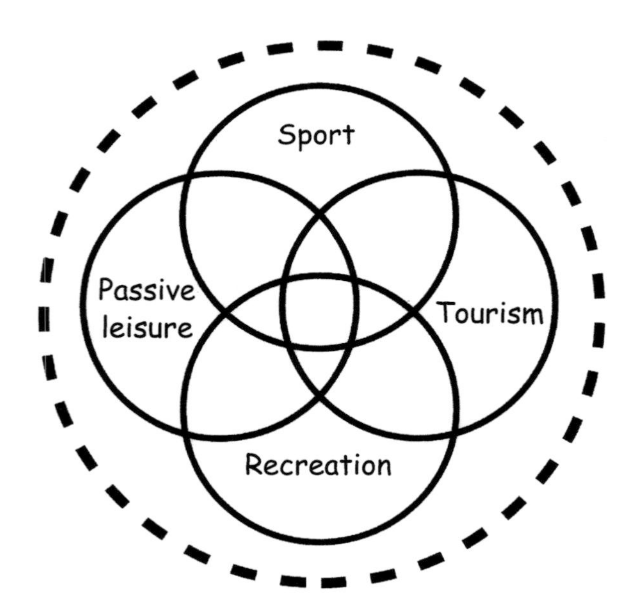

The leisure industry

The leisure industry, including tourism, is the world's biggest single industry and it is still growing at about six per cent per year. Leisure is more than simply things people do in their spare time. The leisure industry covers a wide range of services as the next diagram shows.

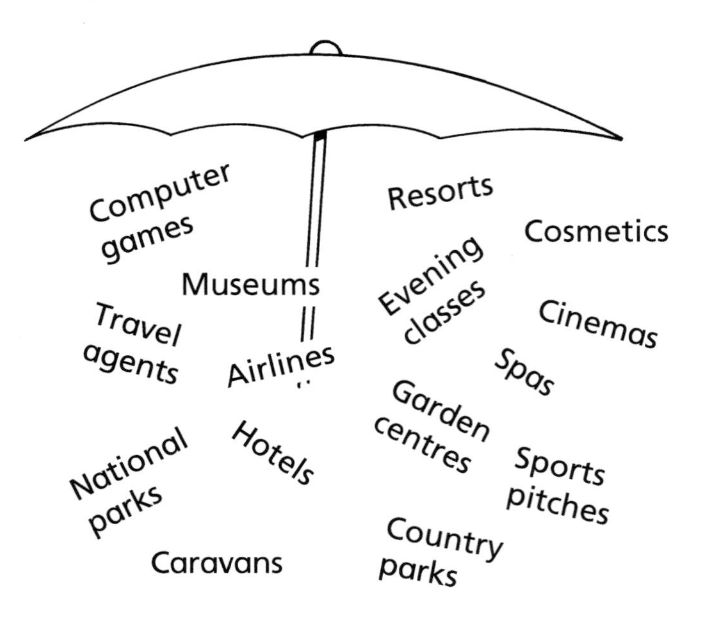

Leisure is about experiences that include ideas of fun, freedom, choice, getting away from it all, feeling good and personal challenge. For many people in the world's richer countries, leisure has become an important part of their lifestyle.

Social and cultural aspects of leisure

Two key factors affect how people enjoy their leisure:

1. People's income and social aspirations (what they can afford and what they want to do).
2. Fashion – for example, in 1984 there were very few mountain bikes in the UK, but now there are thousands.

Planning for leisure

In order to plan what leisure facilities to develop and how many to build, local planners have to think about four different types of consumers.

The obsessives

These are a relatively small number of people who need specific resources, such as the competition swimmer who needs to use a pool for several hours each day.

The leisure rich
These are the large numbers of people who want to participate in leisure activities frequently. An example would be a professional couple who belong to several clubs and societies, such as a theatre group, a squash club or a health club and who go on holiday at least twice a year.

The dabblers
These people are occasional users of a wide range of different resources. For example, a family who might go on a weekend visit to a country park, go shopping, play pitch and putt or go to the cinema. They are difficult to plan for because they may make one visit to a leisure facility and never return, or they may return every week.

The leisure poor
These are a group of people who are infrequent users of a narrow range of leisure facilities. Single parents with young children may be in this group of people who often have severe constraints of time, money and mobility. They may have few opportunities for holidays and, again, are difficult to plan for in relation to how often they use leisure facilities.

Popular leisure activities

The most popular UK leisure activities in 2000 for both males and females aged 16–24 were (in descending order of popularity):

- visiting a pub
- visiting a disco or nightclub
- visiting a cinema
- having a meal in a restaurant
- visiting a library
- taking part in/attending a sports event
- visiting a funfair
- visiting a museum or art gallery
- visiting a theatre.

For people aged 35–44 the list was:

- visiting a pub
- having a meal in a restaurant
- visiting a library
- taking part in/attending a sports event
- visiting an historic building
- visiting a cinema
- visiting a funfair
- visiting a theatre.

Leisure time

Geography objectives (Unit 1)
- To learn about the need for leisure activities and the types of facilities available.

Resources

- Photographs of leisure centres and services that are available locally, such as tennis courts, football/hockey pitches and cinemas (or pictures of similar centres)
- Generic sheets 1–3 (pages 76–78)
- Activity sheets 1–3 (pages 79–81)
- A large-scale map or plan of the local area

Starting points: *whole class*

Tell the children that they are going to look at things people do in their spare time or leisure time. One at a time, show them the pictures of local leisure facilities, including those from Generic sheet 1 on an OHP or in an enlarged version.

Ask questions such as:

- What is this place?
- Why do people like going there?
- Is there a place like this near here?
- If so, where is it?
- What sort of people go to this place – for example, how old, how energetic?

Show the children Generic sheet 2. Tell them that you are going to sort the pictures in some way. Sort them into two groups – activities that take place mostly in summer, and activities that take place mostly in winter. Challenge the children to guess how they have been sorted. Point out that there are some activities that could take place in either winter or summer, such as taking photographs, going swimming and playing on a computer. You could then sort the pictures into sets and sub-sets – one for winter, one for summer and a sub-set for both.

Ask the children if the pictures could be sorted in other ways (indoors/outdoors or need equipment/ don't need equipment).

Tell the children that they are going to think about how they spend their leisure time. Remind them of the activities in the pictures on the two generic sheets and, if necessary, give them copies of these sheets.

Group activities

Activity sheet 1
This sheet is aimed at children who need more support. They have to tick things that they do in their spare time. Then they have to draw pictures of two things they do in their leisure time. Stress that these should be different from the activities shown on the top half of the sheet. Ask the children where these activities take place. They have to write the street or place name (this may be revision of their home address).

Activity sheet 2
This sheet is aimed at children who can work independently. They have to tick boxes to indicate what they do in their spare time. They have to draw three pictures and write the places where these activities take place.

Activity sheet 3
This sheet is for more able children. They have to complete a chart showing how often they participate in different leisure activities. They then have to draw pictures of four different leisure activities and give each a location.

Plenary session

Share the responses to the activity sheets and hold a class census on which are the five most popular leisure pursuits. Ask the children if the answers would be different if they asked their parents/carers.

Pin up a large-scale plan of your area (or project it on an OHP) and ask the children to locate some of the local leisure facilities, such as the football pitch, cinema, swimming pool and shops.

Ideas for support

Make a collection of photographs from travel magazines to include general leisure activities such as swimming, sunbathing, water skiing, football and tennis. Show these to the children and ask them what the pictures have in common (they are all things people do in their spare time).

List the things that many children like to do in their spare time (such as those on Generic sheet 2) and illustrate them with photographs and pictures cut from magazines. Make them into a wall display for the children to refer to when they start the activity sheets.

On a wall display put headings that divide a weekend into mornings, afternoons and evenings and ask the children to put pictures under the correct heading. Some pictures will apply to more than one heading (set).

Ideas for extension

Give the children a copy of Generic sheet 3. Ask them to imagine they live in the city shown in the centre of the map and ask them to plan:

- a trip for half a day in the summer from the city shown on the map;
- a trip for a week in winter from the city.

In each case, ask them to use the map to describe where they would go and why they would choose to go there. Ask them to say how they would get there and where they would stay.

Linked ICT activities

Talk to the children about different types of leisure activities and how some leisure centres/swimming pools may want you to become a member before you are able to use the facilities. Discuss what is meant by membership and show them some examples of what membership cards look like. Find out from if they are members of any clubs or activities. Many may be members of swimming clubs or Cubs and Brownies.

Ask them to work in pairs to take a digital photograph of each other. Tell them that they are going to make their own membership card for any fictitious club that they would like to be a member of. Help them to load the image from the camera onto the writing page and show them how to add their text above and below the picture, changing the font size, style and colour. Print off their final card and cover it in plastic to look like their own membership card.

Leisure activities

Swimming pool

Leisure centre

Aerobics centre

Indoor tennis centre

Self-defence centre

Library

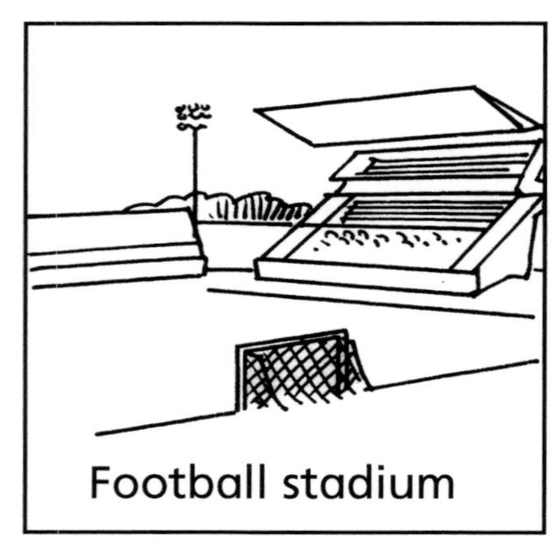
Football stadium

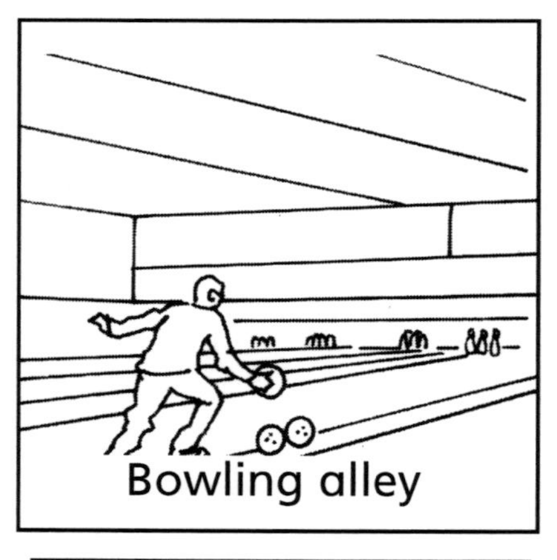
Bowling alley

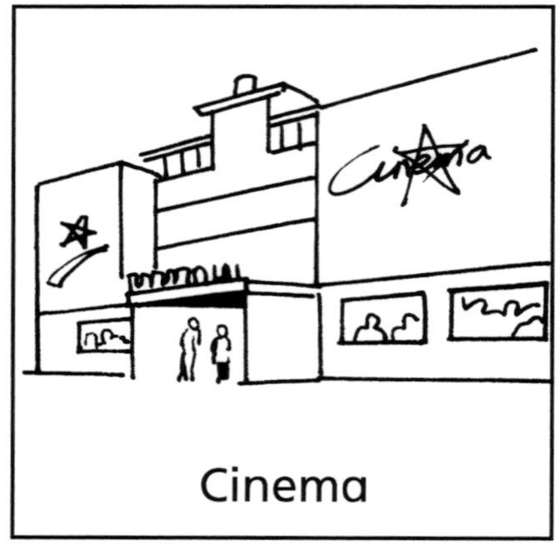
Cinema

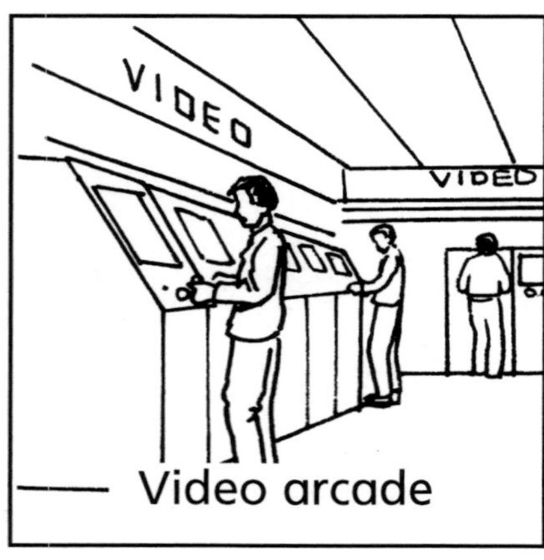

Video arcade

Shopping centre

Horse riding centre

Leisure activities

Playing cricket

Ice skating

Skiing

Building a snowman

Sunbathing

Taking photos

Throwing snowballs

Swimming

Riding a bike

Playing badminton

Playing on a computer

At a barbecue

Leisure activities

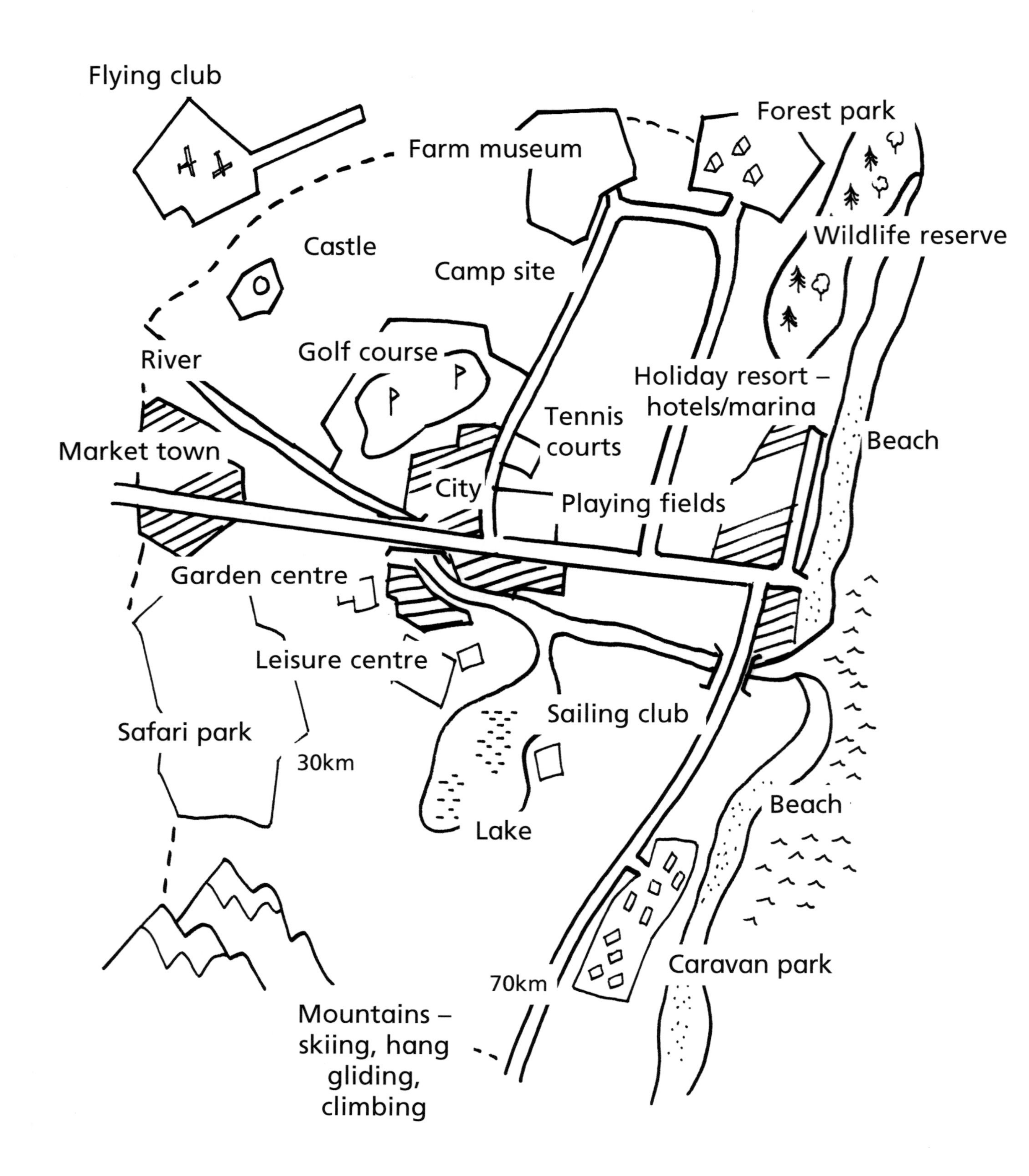

Name ______________________

Leisure time

Put a tick by the things that you like to do in your spare time. 

Reading a book ☐	Going for fast food ☐	Swimming ☐
Playing a video game ☐	Going for a walk ☐	Playing football ☐
Going to the cinema ☐	Cycling ☐	Playing on swings ☐

Draw two other things that you like to do in your spare time:

Write the name of the street where you do each of these things:

Name ______________________________

Leisure time

Put a tick by the things that you do in your spare time. ✓

Go swimming	☐	Talk with friends	☐
Go cycling	☐	Go for a run	☐
Play football	☐	Go hiking	☐
Read a book	☐	Go to the park	☐
Watch television	☐	Go to fast food outlet	☐
Play a video game	☐	Play badminton	☐
Play tennis	☐	Sleep	☐
Watch a football match	☐	Phone friends	☐
Go to the cinema	☐	Sail a model boat	☐
Go to a video arcade	☐	Go windsurfing	☐
Go shopping	☐	Go canoeing	☐
Go horse riding	☐	Go to a place of worship	☐
Watch a video	☐	Go to a disco	☐
Visit a play area	☐	Read a comic	☐

Draw three other things you like to do in your spare time.

On the back of this sheet, write the names of the places where you do each of these three things.

Name ______________________________

Leisure time

Complete the chart to show the things you do in your spare time and how often you do them.

What I do	Less than once a week	Once or twice a week	Three or four times a week	Every day
swim				
cycle				
play football				
read a book				
watch television				
play a video game				
play tennis				
watch football				
go to cinema				
go to video arcade				
go shopping				
go horse riding				
watch a video				
visit a play area				
talk with friends				
go walking				
go to fast food outlet				
sleep				
phone friends				
place of worship				
read a comic				

How often I do them

On the back of this sheet, draw four other things you like to do in your spare time. Write the name of each place where you do these things.

Change

TEACHERS' NOTES

Children need to understand how and why things change and how to anticipate some of the effects of change.

Types of change

It is useful for young children to think about the types of changes that are taking place around them. They might think about:

- short-term changes – in a day;
- medium-term changes – in a month;
- long-term changes – in a year.

Short-term changes
Changes in the weather during a day, or from day to day, are good examples of short-term change. Children need to be helped to think about the implications of these changes for their lives. For example, they might think about how they change what they wear in relation to the changes in the weather. The importance of being prepared in terms of weather changes can be stressed here and the frequent need to take a range of clothing on a journey in order to cope with possible rain, snow, wind or sunshine. The children need to think first about what changes may take place, then about how the changes are likely to have an impact on their lives for good or ill. This is all part of helping to develop their thinking skills and to equip them to function effectively in the world.

Medium-term changes
These changes are things such as children growing up and the ways in which they grow and develop. Again, the important focus is on helping them to anticipate the different types of change and to be able to imagine how they will affect them and their lives. In this context, it may also be useful to talk about changes in the area around the school. Ask children to identify some of the changes that they may have seen in and around school. These might include things such as new classrooms or redecorated classroom areas.

More widely, there may be new houses being built locally or a new road or simply changes to the local bus routes. In each case it is important to help children to think first about why the change has taken place and, perhaps, why it was necessary, then how things are likely to change as a result of this development, followed by thinking what all this means for them. In this way, the children will begin to understand that they are able to have an influence on change as well as having to react to it. Children should not be led to believe that they are at the mercy of change and are powerless to bring about improvements to change.

Again, in this context it may be useful to consider any debate in the local press about the desirability or otherwise of a change, such as opening a new supermarket, widening a road, opening a new fast food outlet or banning parents from parking too close to the school gates. This provides a good opportunity to show the children that there are often several views about a particular change and that these views may conflict, with some people being in favour of a change and others against it.

Long-term changes
These are changes that may take a year or longer. They might include the development of a new housing estate or the demolition of an old building or some old houses and factories followed by redevelopment of the site. Many children find it harder to think about long-term changes than about short-term ones because, by definition, the whole process takes so much longer and if it is in their local area they will assimilate the change without necessarily realising its implications until it is too late. For example, the closure of a local swimming baths might be discussed for a long time, but it is only when the baths are closed that the children (and many adults) will begin to think about the true impact of the changes. This is part of helping children to imagine what the impact of a proposed change will be and it is a key element of this chapter.

Organising the children into groups and encouraging them to brainstorm what they think may be the likely effects of a particular change will help all children to develop their thinking skills.

Change

Geography objectives (Unit 1)
- To learn how places change for better or worse over time.

Resources

- A copy of *Window* by Jeannie Baker, published by Red Fox, 1991
- Generic sheets 1 and 2 (pages 85 and 86)
- Activity sheets 1–3 (pages 87–89)

Starting points: *whole class*

Tell the children that they are going to look at how places change over time. Show them the pictures of a high street before and after redevelopment on Generic sheets 1 and 2 on an OHP or in enlarged versions.

Ask the children to describe the main changes between these two pictures. These are:

- pedestrianisation – no lorries unloading on the street and no cars and vans to knock people down and cause air pollution. This involves the removal of road signs which changes the appearance of the area;
- removal of overhead wires – to make the area look more attractive;
- planting trees and adding plant containers – to improve the appearance of the area and to attract wildlife.

Ask them to point out all the changes they can spot. Talk about each of the changes. Ask:

- Why do you think this change was made?
- Do you think the changes have made the area better or worse?

If you have a copy, show the children the book *Window* by Jeannie Baker. Explain to them that they have to make up the story to the book, because the book contains only pictures. These pictures show the view from a window in a house. They also tell the story of a child growing up into an adult. The view from the window changes over the years as the trees are cut down and new houses and roads are built over the whole area.

You may decide to show only some of the pictures to give the idea of how places change. Ask questions such as:

- How do we know how old the baby is now? (See birthday cards at 2, 16, 20 and 24.)
- What can you see out of the window?
- What is this area like?
- What birds and animals can you see?

In each case, when you look at a fresh picture, ask:

- What has changed?
- What has stayed the same?

Encourage the children to describe all the changes they see. Ask them:

- Who do you think will like these changes? (People looking for new homes.)
- Who will not like these changes? (Wildlife, locals.)

Tell the children that they are now going to think about changes in places in the UK. Remind them of the changes seen in the generic sheets.

Group activities

Activity sheet 1
This sheet is aimed at children who need more support. They may have some difficulty in identifying features that have changed. Explain that the pictures show the same street in 1860 and in 2001 and that they have to spot the changes. Once they have spotted the main changes and coloured them in (blacksmith/stables and shop to garage; house to superstore; pond and woodland to road; big house and park to car park and factory/office blocks; woods to multi-storey car park; horse transport to cars, buses and lorries), they have to write three sentences about one change on the back of the sheet. They then have to draw a change in their local area, such as a new shop that has opened.

Activity sheet 2
This sheet is aimed at children who can work independently. They have to spot a number of changes in the development of an area between 1860 and 2001 (see the notes for Activity sheet 1 on page 83). They have to number each change and describe it on the back of the sheet. Some children should also be able to suggest reasons for the changes they have identified.

Activity sheet 3
This sheet is for more able children. They have to identify and list the main changes in the street between 1860 and 1900 (time to get to the town, mode of transport to town, number of houses in the street) and between 1900 and 2001 (time and mode of transport to town, shops and office blocks replacing most houses, street lighting, no trees, hedges or fields). Then they have to draw how the street might look in 2020. This might need some teacher input in terms of questions such as:

- Will buildings be taller?
- Which old buildings might be pulled down?
- What new buildings might be needed?
- What might shops look like?
- What might offices look like?
- How might you travel down the street?

Plenary session

Share the responses to the activity sheets and discuss what things have changed (houses, roads, traffic, open space, woodland) and what things have stayed the same. Talk about how we decide who has benefited from the changes and who has lost out.

Ideas for support

Ask the children to study a small part of Generic sheet 1 and Generic sheet 2 at a time, starting with the front, then the middle section, then the skyline and rear parts of the pictures. In each case ask them to say how those parts of the pictures have changed (front: no traffic, no traffic sign, seats, trees where road was; middle: no parked cars, people can walk in road, easy to cross the road; back: no overhead wires, new street lights).

Ask the children to bring in photographs of themselves as babies. Ask them also to bring in a more recent photograph of themselves. Use the baby photos to ask them to identify each other. Ask how they know who the photo shows and why it is difficult to tell.

Next, ask the children to compare a recent photograph of another child with the baby photograph of the same child. Ask them to list all the ways in which they can see that the child has changed – for example, height, weight, hair colour, eye colour, and so on. Stress that things are changing all the time, not just in people but in the environment around us. For example, point out any changes in the school, such as a new teacher, a new window, or an area that has been repainted.

Ideas for extension

Select an area near the school with which the children are familiar. This could be the school playing fields, an area of open space or an area of derelict land. Ask the children to study this area.

- How big is it? (They can measure it or use a large-scale simple plan.)
- What is it like? (Flat, steep, grassland, wooded?)

Then ask them to work in groups of four and to suggest ways in which the area could be developed in the future. They should include drawings of how the area might look, together with reasons why they think their project should go ahead. Alternatively, the children could suggest ways in which the school playground could be made more environmentally friendly.

Linked ICT activities

Discuss the changes that computers have made to people's lives. Children at this age will know only a world in which computers play an integral part but it is important for them to understand that computers have not always been around. Discuss the places where they see computers being used and what they think they are used for. Talk about how they think jobs could be carried out today without the use of computers. What do we use computers for and how would our lives change if we didn't have them? Take the opportunity to talk about the different jobs people in school and in the locality of the school have. Do any of these people use technology to help them to do their job?

Young children often don't realise that most of the things we use daily use technology – for example, videos, washing machines and cars. Use pictures of old and new everyday items and talk to the children about how these would have been used in the past and how they are used today.

Change

Before

Change

After

Name ______________________________

Change

Look at these two pictures of the same street.
On the second picture, colour in red everything that has changed from the first picture.

A street in 1860

The same street in 2001

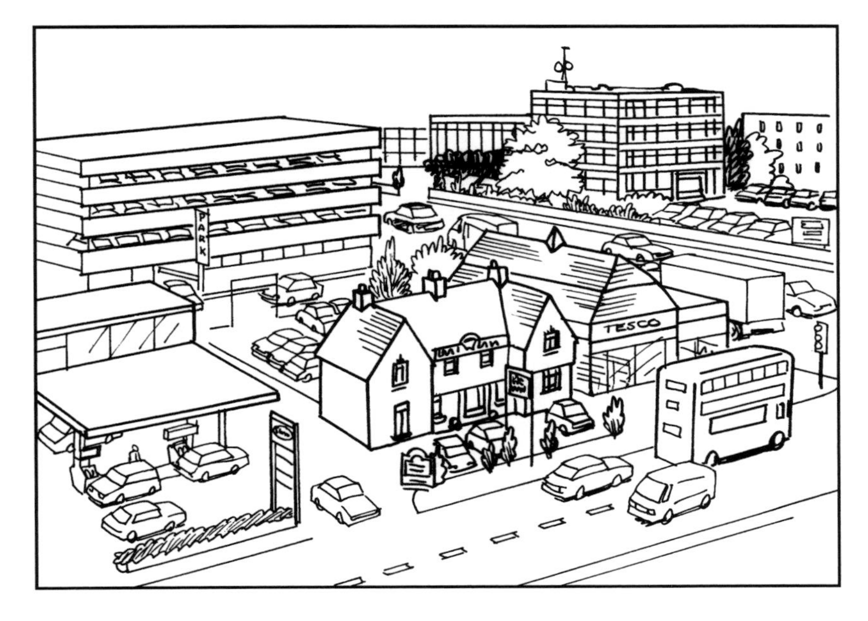

On the back of this sheet, write three sentences about one of these changes. Then draw a change that has happened in your local area.

Name ______________________________

Change

Look at these two pictures of the same street.
Circle all the changes between the pictures and give each one a number.

1860

2001

On the back of this sheet, write a sentence for each number to describe the change.

Name ______________________________

Change

These pictures show how this street has changed. On another sheet of paper, list all the changes you can see between:

a) 1860 and 1900
b) 1900 and 2001.

In the box at the bottom of the page, draw what you think the street will look like in 2010.

1860

Walk to town takes 50 minutes.

1900

Journey to town centre by omnibus takes 15 minutes.

2001

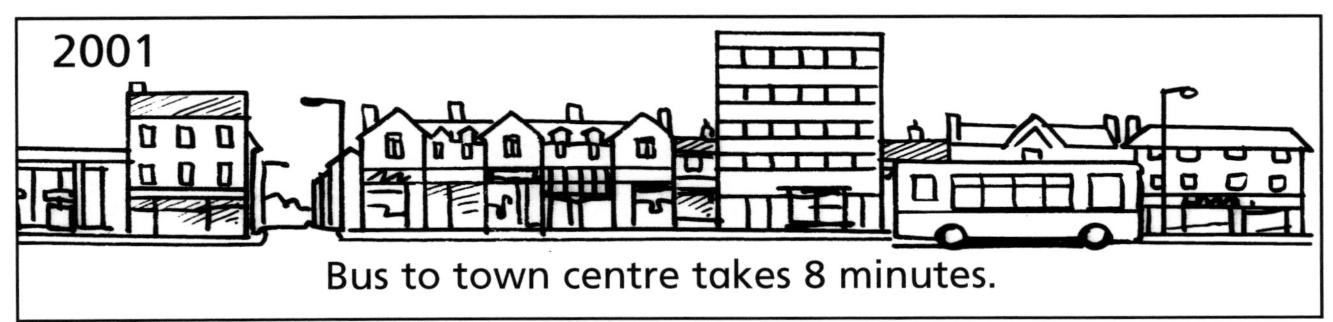

Bus to town centre takes 8 minutes.

2020

Journey to town centre by __________ takes ____________________

Roads

TEACHERS' NOTES

Looking at local roads and local traffic is an opportunity to develop ideas about safety.

Background to roads

The Romans built the first proper roads in Britain almost 2,000 years ago. The roads they built were mostly straight and had a good surface of stone. For at least 1,000 years after the Romans left, most new roads in Britain followed winding routes. This was to avoid steep slopes, marshy areas and dense woods. Some roads were built along valley sides. This was so the river valleys could be used as natural routeways for transport.

Then, about 300 years ago, people began to travel in stagecoaches. The roads needed improvement and maintenance to make travel easier. To pay for the roads, travellers were charged a fee called a 'toll', rather like we now pay car tax. The toll was collected at a tollhouse beside the road and old tollhouses can still be seen along some old main roads. Today toll roads are coming back into fashion as a way to solve the problems of too many vehicles on too few roads.

Roads have developed a lot since the time of stagecoaches and tollhouses. The developed world now has an extensive network of roads and motorways. Roads are such a vital part of everyday life that people tend to take them for granted, until the road falls into disrepair or becomes very busy. Children need to be introduced to the key ideas of what a road is:

- an area with space for vehicles and places for people (pavements); and
- the two groups are kept separate on purpose.

Children also need to be introduced to the idea of traffic; what it is and how it changes.

One way to approach a study of roads is to talk about the advantages of travelling on longer journeys by road (as opposed to rail or water). The main advantage is convenience: door-to-door journeys without having to change types of transport. Cars are also being used more and more for journeys to work, for shopping and local recreation. This is because they are relatively cheap to use over short distances. Buses struggle to compete with cars on cost and convenience. Lorries have become much more specialised in what they carry, such as bulk flour or oil, or containers. Goods loaded on a lorry in the UK can be sent anywhere in Europe without having to be transferred to another vehicle. The lorry can drive along roads, get on a ferry to cross water or cross the Channel by Eurotunnel train and then drive along again.

Cross-curricular work on roads

Numeracy

Graphs/percentages/fractions
Traffic counts looking at:
- types of vehicles
- colours of vehicles
- times of day
- days of the week.

Work on sets
- Types of vehicles.
- Colours of vehicles.
- How we travel to school.
- People who help us.

Statistics/map work
- Accident sites around the school.
- Safe crossing places.
- Our houses and homes.
- Routes to school.
- Interpretation of bar charts and other pictorial representations.

Circles
- Using a wheel to measure diameter, circumference and radius.

Fast and slow
- Identification of things that move fast and slowly.
- Understanding of how fast we can move.
- Estimation of speed and distance.
- Reaction time tests.
- Stopping distances of vehicles at different speeds.

Literacy

Creative writing
- The scene of an accident.
- A court case involving a motoring offence.
- The problem of road safety from a driver's point of view.

Drama and role play
- People in the street.
- People who help us.
- Reconstruction of an accident.

Science

Weather
- Changing road surfaces in different seasons.
- Visibility problems for all road users.

The senses
- Stop, look and listen.
- Tape of traffic sounds.
- The importance of using our senses when we cross the road.

Camouflage/conspicuousness
- Why animals use camouflage.
- Why we must be conspicuous.
- Looking at clothes in relation to the weather and senses – for example, hoods on coats impair visibility and hearing.

Friction
- Tyre treads and why we need them.
- Various types of road surface.

History
- Roads and their development.
- The development of the bicycle and the motor vehicle.
- Events in history.
- Travel, past and present.

Art

Colour and shape
- Circles/wheels.
- Designing road signs.
- Red, amber, green.
- Abstract painting/drawing.

Modelling
- Junk models.

Geography
- Identification of street furniture.
- Identification of safe crossing places.
- Location of signs.
- Trails around the school.
- Traffic counts.
- Pollution hot spots.

Roads

Geography objectives (Unit 2)
- To learn about the character of a place.
- To ask geographical questions and to use geographical terms.

Resources

- Photographs of busy local roads showing a variety of traffic – lorries, buses, cars, bicycles
- Generic sheets 1 and 2 (pages 94 and 95)
- Activity sheets 1–3 (pages 96–98)

Starting points: *whole class*

Explain to the children that they are going to look at local roads and traffic. In particular, they are going to look at busy roads and quiet roads. Show them the picture on Generic sheet 1. Ask questions such as 'What is traffic?' and let them explain the variety of vehicles that make up traffic.

Encourage them to talk about the details of the traffic and, in particular, to deal with different sizes and shapes of vehicles as well as the range of sounds different vehicles make. Encourage some imaginative thinking with questions such as:

- Where do you think the lorry might be going?
- What do you think it might be carrying?

Next, show the children some enlarged pictures from Generic sheet 2. Tell them that you are going to sort the pictures into two groups. Sort the images into those with two wheels and those with four or more wheels. Share ideas on how things within the groups are different.

Ask them to suggest other ways in which the pictures could be sorted, such as large and small, slow and fast, heavy and light. They could also place the pictures in size order and attach them to the board.

Tell the children that they are now going to study the traffic on two roads – Shaw Lane and Green Road. (Alternatively, choose local names for a busy road and a quiet road.) Explain that some children have counted the traffic on these roads in a survey and the class are going to study the results.

Group activities

Activity sheet 1
This sheet is aimed at children who need more support. They can recognise the main types of traffic and are familiar with the use of symbols to represent each vehicle. They have to colour in each type of vehicle in a different colour and find the totals for each type. When they have done this for both roads, ask:

- Which is the quieter road?
- How do we know this?

Then they have to think about the relative speeds of different things and colour the outer rings in red for fast-moving things and green for slow-moving things. This can be linked to safety, especially the need to take care with fast traffic and busy roads.

Activity sheet 2
This sheet is aimed at children who can work independently. They are familiar (or can become so quickly) with tally charts and the names of the different elements of traffic. They have to calculate the total numbers for each vehicle type on each street, then use these to create block graphs for each street.

Activity sheet 3
This sheet is aimed at more able children. They are familiar with tally methods of vehicle counts. They have to calculate the numbers of each vehicle type and the overall totals for each road. They then have to use data to draw a graph of traffic in each street.

Plenary session

Share the results of the activity sheets and decide which was the quiet street and which was the busy street. Ask the children to name (other) busy streets near the school, or near their homes, and (other) quiet streets in the local area. Recap the key terms – traffic, survey, busy, quiet.

Ideas for support

In approaching Activity sheet 1 it is best to ask the children to do the cars in Shaw Lane, then the cars in Green Road, as a starting point. Then move to the lorries in each of the two streets, and so on down the list.

It is also useful to talk about each of the items shown in the circles, because they might not immediately recognise a snail, nor think about how fast a young person might move. Emphasise that they are thinking about how fast or slow each of the things will move.

Provide the children with a series of toy vehicles, from cars, vans and lorries to buses and bicycles. Let them play with the vehicles and then ask them to sort them into two or more groups. Let them decide the groups, which could be on the basis of colour, size, shape or speed.

Take your own photographs of traffic, stick them on card and then cut them in half. Shuffle the cards and ask the children to fit the pieces together again. With older children, the cards could be cut into three or four pieces to increase the level of difficulty.

Use Generic sheet 2 to create a set of traffic cards to play a game. Give each child one or two cards. Then say, for example, 'Hold up a bus.' The children with the cards hold them up. Can they say what is different about the two types of buses on the pictures? Which one might make the most noise? Why?

Ideas for extension

Record a range of traffic sounds, together with other sounds, such as rivers or falling water. Challenge the children to distinguish the traffic noises from the other sounds. Alternatively, you could ask them to match the pictures from Generic sheet 2 with the sounds.

Ask the children to think about noisy places and quiet places along a stretch of road they visit or know. Ask them to explain why some areas are so noisy.

- Is it because traffic is stopping, or pulling away?
- Is it because the traffic is forced to go very slowly?
- Is it noisy near bus stops?

Ask why other places are quieter.

- Is there less traffic?
- Are there fewer bus stops?
- Are there traffic lights or pedestrian crossings?

Linked ICT activities

Talk to the children about how dangerous traffic can be and what they think the dangers are. Make a list with them to show how traffic could be dangerous – for example, driving too fast, parking in the wrong places, too many vehicles in pedestrian areas and so on.

Talk about keeping safe when they are around any traffic, whether it is fast-moving or stationary. Work with the whole class to put together a class set of road rules. Tell the children that you are going to do the first three rules together and that they are then going to put five of their own rules together, working in pairs. When everyone has had the opportunity to think of some of their own rules, the whole class could then work together on deciding which rules to use to add to the three they started with and create a list of ten in total.

Create a word bank of useful words, which the children may need in order to complete their five road rules. Use a simple word processing program such as *Talking Write Away* or *Textease*, both of which allow you the option of creating word banks. Working in pairs, give the children the opportunity to write their set of five road rules, saving their work as they complete each rule and printing out the final version. When they have completed their work and saved it, ask them to read out their own rules (there will probably be a lot of repetition within the rules). Using the computer or flip chart start to pick out the key rules which the children have identified to add to the three that you have already written.

A busy street

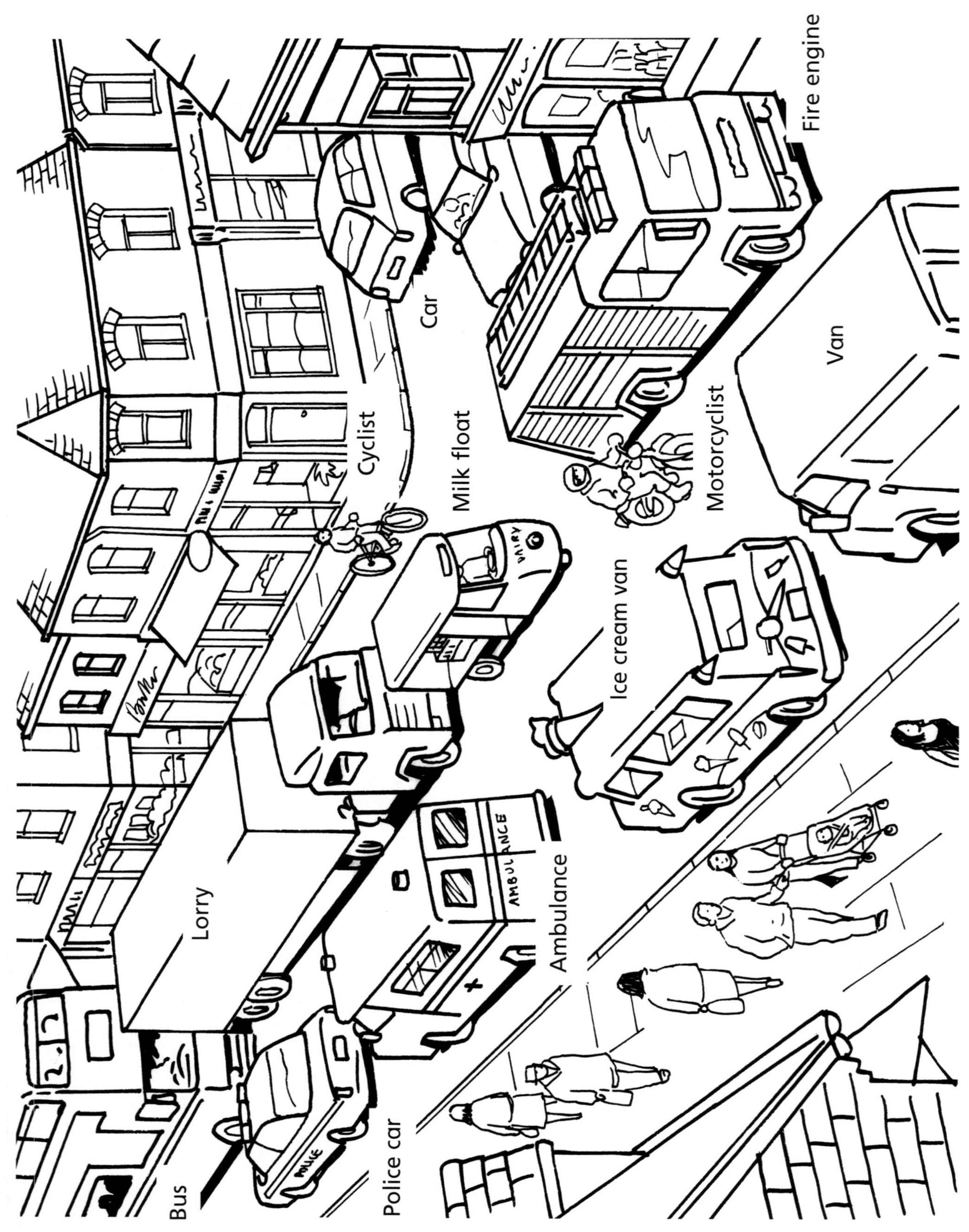

Roads

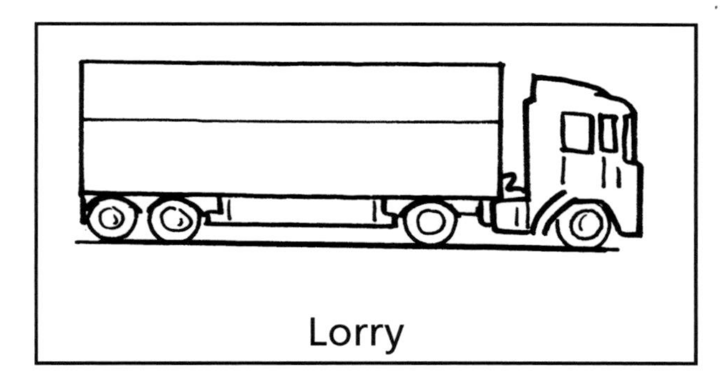

Lorry

Tanker lorry

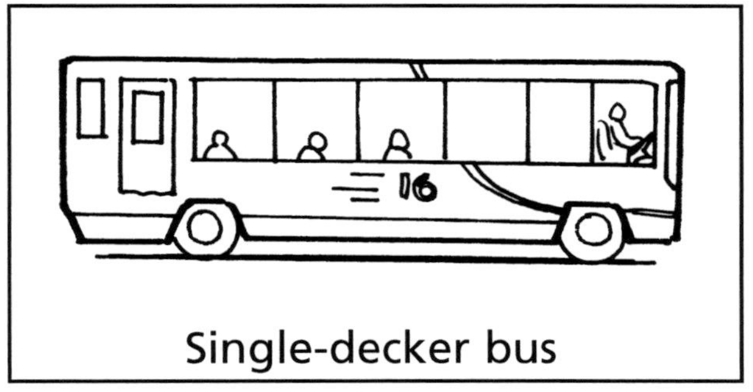

Single-decker bus

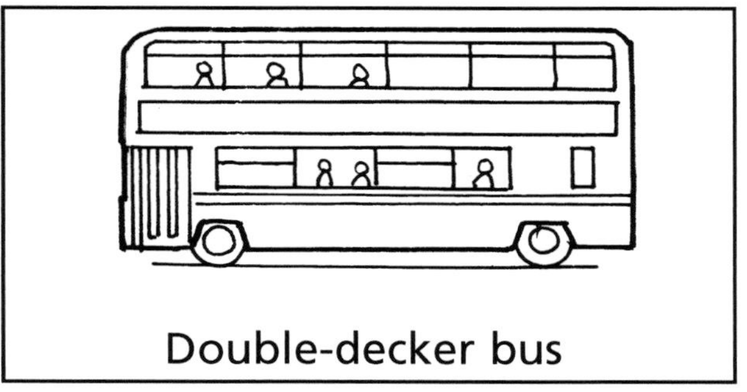

Double-decker bus

Small car

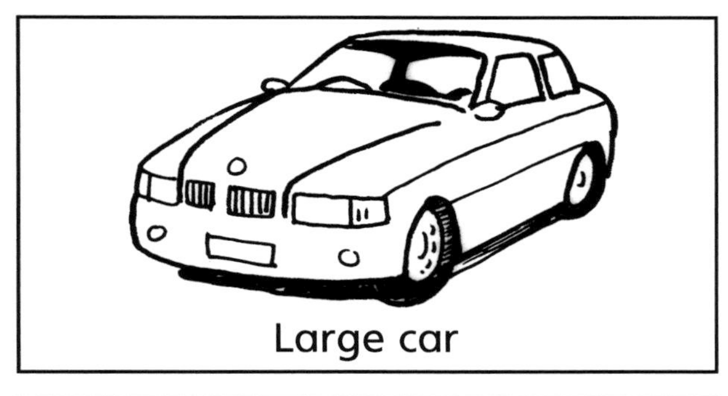

Large car

Shopping bicycle

Mountain bike

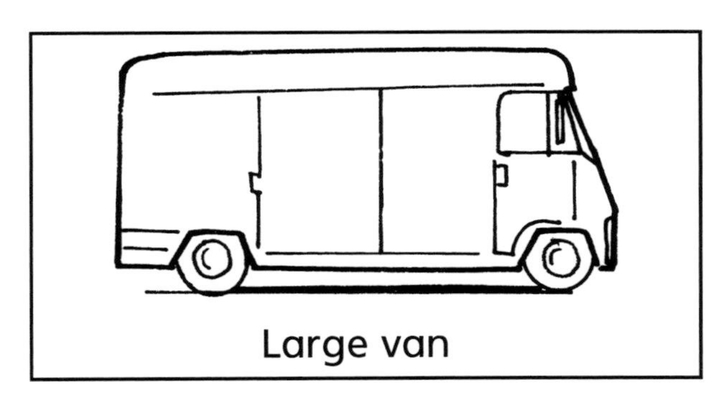

Large van

Small van

Name ______________________________

Roads

Colour each **type** of vehicle a different colour.
Find the total for each type of vehicle on each street.

Shaw Lane

Vehicle		TOTAL
Car		
Lorry		
Van		
Bus		
Motorcycle		
Bicycle		

Green Road

Vehicle		TOTAL
Car		
Lorry		
Van		
Bus		
Motorcycle		
Bicycle		

Colour red the circles around the fast things below.
Colour green the circles around the slow things below.

Name ______________________________

Roads

Calculate the totals for each **type** of vehicle on each street.

Shaw Lane

Vehicle	Tally	TOTAL
Car	𝍸 𝍸	
Lorry	𝍸 III	
Van	𝍸 𝍸	
Bus	𝍸	
Motorcycle	𝍸	
Bicycle	IIII	

Green Road

Vehicle	Tally	TOTAL
Car	𝍸 II	
Lorry	I	
Van	IIII	
Bus	I	
Motorcycle	II	
Bicycle	III	

Make a block graph for each street.

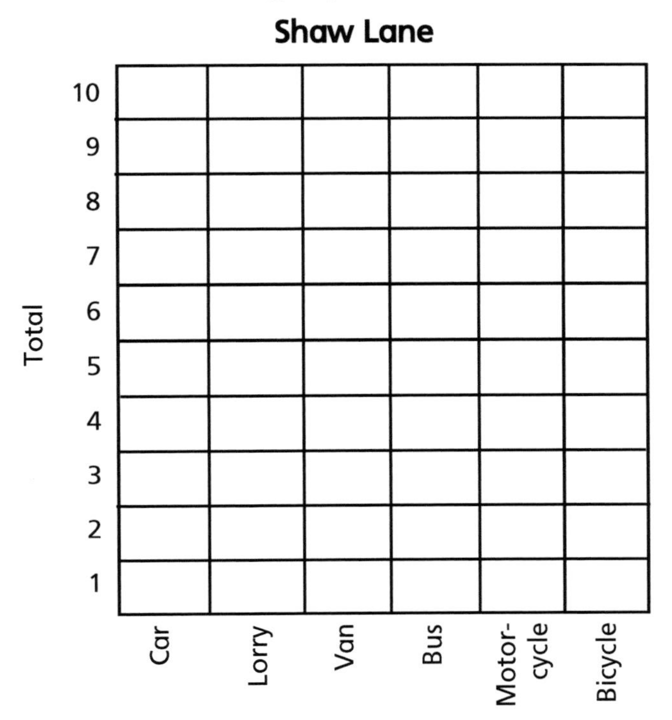

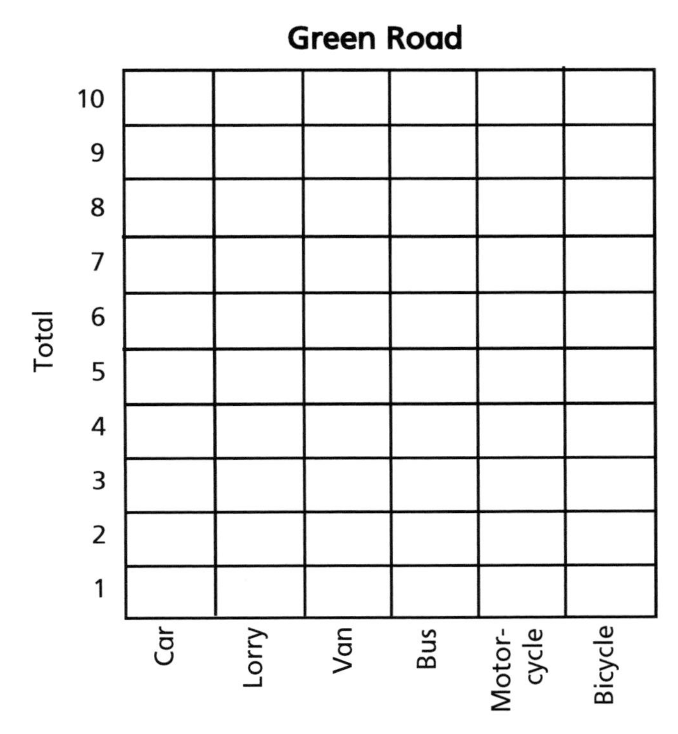

Name ______________________

Roads

3 ACTIVITY SHEET

Calculate the totals for each type of vehicle on each street.

Shaw Lane

		TOTAL
Car	卌 卌 卌 卌 IIII	
Lorry	卌 卌 II	
Van	卌 卌 IIII	
Bus	IIII	
Motor-cycle	卌 I	
	Overall total	

Green Road

		TOTAL
Car	卌 I	
Lorry	I	
Van	卌	
Bus	III	
Motor-cycle	II	
	Overall total	

Now draw a graph to show the traffic in each street.

Shaw Lane **Green Road**

Parking

CHAPTER 10

TEACHERS' NOTES

There are about 24 million cars in the UK and a quarter of all families own two cars. Some families even have three cars. There are also about half a million lorries, buses and other types of vehicles and they have all got to park somewhere.

People travelling into towns or cities for shopping, business or leisure need somewhere to park their car and this takes up space. People will park either in designated car parks or at the side of the road. Roadside parking can be dangerous because it blocks the view for people who want to cross the road; it also reduces the width of the road, which can cause problems for lorries and large vans delivering or collecting goods from shops or businesses.

Long queues of cars waiting to get into a car park can cause serious congestion problems and have a damaging impact on the environment with stationary cars pumping out exhaust fumes.

Another issue is the cost of parking in towns and cities. Because there is limited space to build car parks and the cost of the land is high, car park owners pass this cost on to the car driver in high parking charges. On the other hand, car owners can always go and use an out-of-town shopping centre where they can park for free. This has a dramatic effect on businesses and shops based in the middle of towns and cities and contributes to many small shops and businesses having to close down.

Many towns and cities have introduced 'park and ride' schemes. The idea is that people needing to go into a town or city where there is limited parking space, can leave their car in a very large car park and then travel to and from the town centre on a special bus. The combined cost of parking and bus fare is very much less than that town's central parking cost. Several cities have more than one park and ride, so people can use the one closest to home.

Parking outside schools is a major hazard for children. A crossing warden, pedestrian crossing, double yellow lines and school entrace markings are the main ways of protecting them coming in and out of school. Urban clearway signs and markings provide for no stopping during the times shown on the sign except for setting down and picking up passengers.

Other measures slow traffic outside schools and may include speed bumps and 'school safety zone' speed limits of 20mph.

Drawing mental maps

The map work activities in this chapter develop logically from those in the earlier chapters, especially Chapters 1 and 3. The QCA scheme of work for Unit 2, 'How can we make our local area safer?', suggests that children should be encouraged to 'visit the road outside the school and ... either to record on a map the various ways used to control traffic, or to observe the parking controls and then to return to the classroom to draw a map from memory to show their observations.'

This is quite a difficult task to ask children to undertake unless they have already completed the activities described in the teachers' notes for Chapter 3. Continue and develop work on the plan view of objects. Give particular attention to developing the idea of drawing a sketch map of the classroom. This can be done quite simply using a plan like the one on Generic sheet 1. Most children find using 1cm squared paper a big help in this activity. Do not worry if their sketch maps include both pictorial and symbolic elements – that is natural at this stage.

Parking and safety

Geography objectives (Unit 2)
- To observe, recognise and describe the main ways in which parking is controlled.
- To undertake simple mapping tasks.

Resources

- Generic sheets 1–5 (pages 103–107)
- The words from the word bank on Generic sheet 4 written out and cut up
- Activity sheets 1–3 (pages 108–110)
- Sticky tape
- 1cm squared paper
- Card and scissors
- Simple map of the area outside the school, showing roads and pavements, large enough for the children to add road signs and so on
- Photographs of parked vehicles in the area near your school to show problems created by parking (for example, narrower road, hard to cross the road)

Starting points: *whole class*

Tell the children that they are going to make a sketch map of the classroom. Let them use 1cm squared paper. Use Generic sheet 1 as an example. Suggest that they draw round a piece of card that you have cut to the correct size to represent a table. Leave out the drawing of chairs at this stage if it is too fiddly. In all the work, emphasise that sketch maps are drawn as if looking down from above. Ask questions such as:

- Which things are nearest the door?
- How many children's tables are there?
- Which desk is nearest the board?
- Which table is nearest the window?

The next step in helping children understand what maps do and do not show is an activity based on the pictures on Generic sheets 2 and 3. There are three views of a school, and a map of the same area. Show the first picture and ask:

- What can you see in this picture?
- Which way is the car park?

Explain that this is the view they would see if they walked into the school.

Now show the second picture. This is a view of the same school as if taken from the top of a nearby block of flats. Ask:

- What can you see now?
- What new things can you see? (Playground, car park, tennis court.)

Move on to the third picture and explain that this is the vertical view: what they would see if they were in a helicopter directly above the school. Again ask:

- What can you see?
- What new things can you see? (Patio area, field, more trees.)
- How does the school look different now? (They can only see the roof and there is no indication of how tall the building is.)

Then show the fourth picture – the map of the school – and explain that it is based on the last picture so it shows the vertical view. Point out the elements on the map that are also shown in the picture – school, tennis courts, river. Also point out things that are not shown – people and vehicles.

Take the children to visit a short section of road outside the school. It may be possible to stand in the playground and see the road. Point out key landmarks to the children, especially road signs and street furniture such as street lamps, post boxes, telephone boxes, litter bins, street name signs or any other key features. In pointing out the features, stress where they are, such as close to the corner, near the school or next to the tree.

Back in class, show the children the simple base map that you have drawn showing the pavement and the main roads outside the school. Ask them to add to this base map by locating road signs, street furniture, pedestrian crossings and road markings that they have seen. Once they have completed this, a brief return trip to the road will help them to check the accuracy of their mapping and memory.

Tell the children that they are going to look at pictures of a street and some of the features found in the street. Show them Generic sheet 4 on an OHP or in an enlarged version. Ask them to label each of the features shown with a box. Let them come out and stick the correct name next to each feature. Stress the importance of the difference between the road and the pavement. Talk about the different things that belong on the road (cars, vans, lorries, buses) and those that belong on the pavement (people, street lights, signs, postbox, telephone box). Emphasise that people belong on the pavement.

Next show Generic sheet 5 either on an OHP or in an enlarged version. Ask questions such as:

- How is this different from the last picture?
- What else can you see?
- How is the road different?

Ask the children to count the number and type of the different parked vehicles: two lorries (including the dustcart), one bus, two cars, one van, one motorcycle.

Ask questions such as:

- Why do you think the lorries are parked there? (To unload or to collect rubbish.)
- Why do you think the cars are parked there? (For people to go to the shops or to go to work.)
- Why are parked vehicles dangerous? (They make it difficult for drivers to see small children or for pedestrians to see traffic on the road, and they make the road narrower creating traffic jams.)

Tell the children that they are now going to look at some of the ways in which parking is controlled to prevent these dangers. Discuss each term on the activity sheets – 'urban clearway', 'pedestrian crossing', 'crossing warden' and 'no parking lines'.

Group activities

Activity sheet 1
This sheet is aimed at children who need more support. They can recognise the four main ways in which parking is controlled – the sheet does name the ways, but it would be useful to go over them first to ensure that they understand each one. They have to colour the no parking lines yellow. On the back of the sheet, they have to design a sign that warns pedestrians not to cross the road between parked cars (such as a red circle with a child walking out between parked cars and a broad red line across it).

Activity sheet 2
This sheet is aimed at children who can work independently. They can visualise an area when seen from above. They have to think about where to locate each of the four features to control parking.

Activity sheet 3
This sheet is aimed at more able children. They are very familiar with the relationship between the picture and the map. They have to consider different possible locations for a pedestrian crossing, then mark it on both the picture and the map and explain the reasons for their decision (on the back of the sheet). They then have to locate a place for a crossing warden and justify this, together with a location for double yellow lines and an urban clearway sign.

Plenary session

Share the responses to the activity sheets. Revise the four main methods of controlling parking. Ask the children:

- Why do so many drivers try to park close to our school gates each morning and afternoon?
- Why is this dangerous?
- How could it be avoided? (By parents walking from home to school with their children, or parking safely a short distance away and walking from there.)

Ideas for support

Talk to the children about each feature, starting with the pedestrian crossing. Ask the children what it is and how we should use it. Do the same with the crossing warden, especially as many children will call this person the lollipop man or lady. Point out the 'no parking' lines on the picture and lastly draw attention to the urban clearway road sign.

Play parking, traffic and people, in PE in the hall. Mark out the pavement and the road with chalk. Use some chairs to represent vehicles parked parallel to the pavement. Let some children be drivers of vehicles moving along the road. Other children should stand on the pavement trying to cross. Encourage them to look for a crossing warden or police officer to stop the traffic and help them to cross. The children can act out an accident when one of them tries to cross the road and is hit by a car, bus or lorry. Some children can arrive as paramedics and take away the injured child. Throughout the activity, stress the dangers of crossing between parked cars, and the need to find a safe place to cross. Ask the children to say what a safe place is – where there are no parked cars, where children can see in both directions and a place away from road junctions and bus stops.

Ideas for extension

Make a study of the area around the school. Suggest at least one, or preferably two, new crossing points close to the school. Ask the children:

- What are the main dangers for children in crossing near the school?
- How would the new crossing points improve things?

Identify places within the school where there are problems with the volume of children moving back and forth at certain times of the day. Design, produce and select locations for a series of signs to be mounted in the school to direct the human traffic, using 'Give way', 'Keep left', 'Stop' and other suitable road signs.

Linked ICT activities

Using a simple graphics program such as *Dazzle* tell the children that they are going to design their own signs to stop people parking in places around the school, which presents a danger to the children. Show the children how to select a font from the program, change its size, format and colour and position it where they want on the screen.

First, ask the children to type in their no parking message on the page and to save it. Show the children some examples of signs and encourage them to change the size of the letters and the colours to make the sign more visible. At this point show the children how to save their work and give their file a name.

Continue with the sign by creating a frame or a border around the text and adding some clipart and changing the background colour of the sign. Encourage the children to save their work as they progress to completing the sign. Print out the final signs and have a class competition to see which is the winning design.

A classroom plan

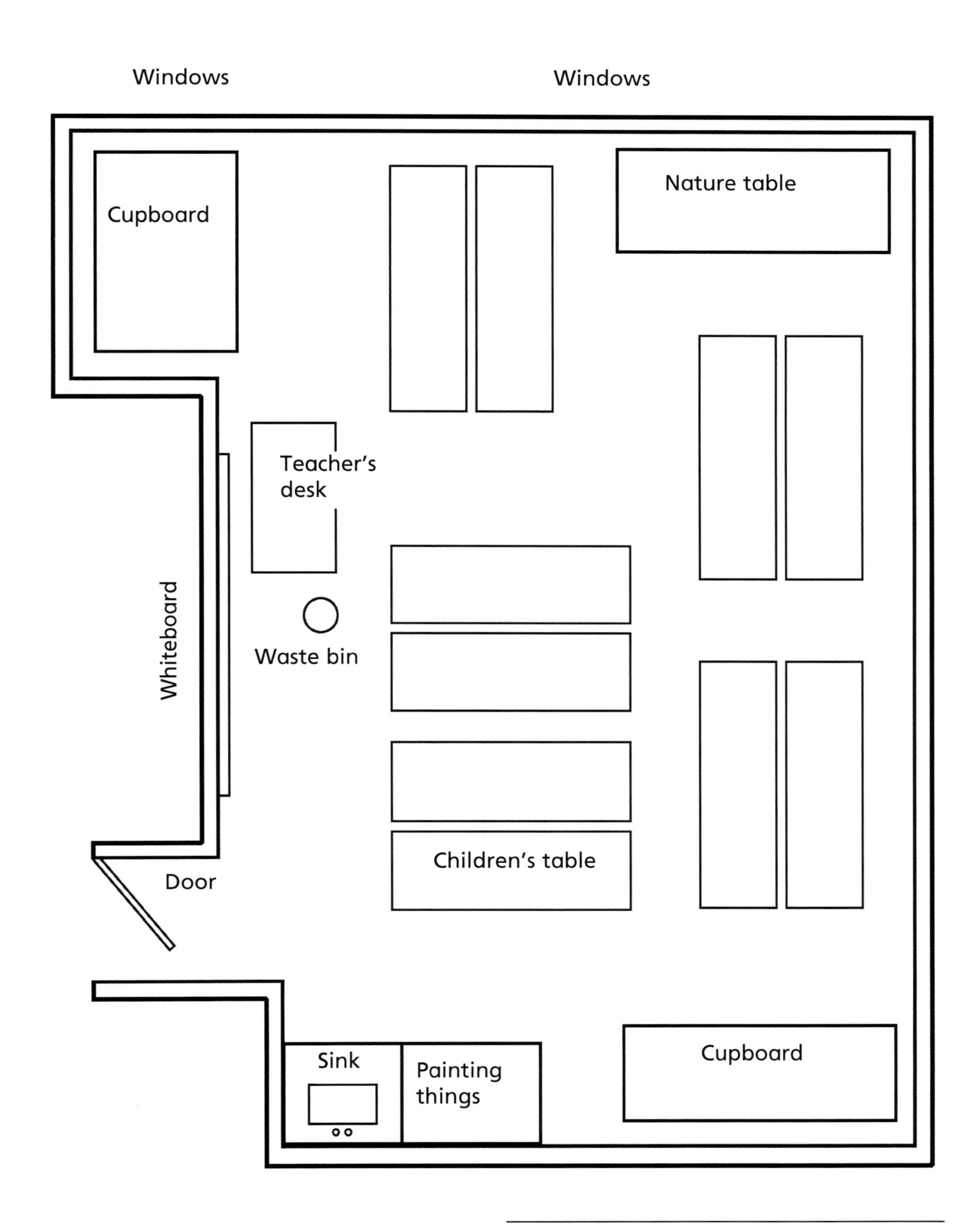

Two different views

1

2

A view and a map

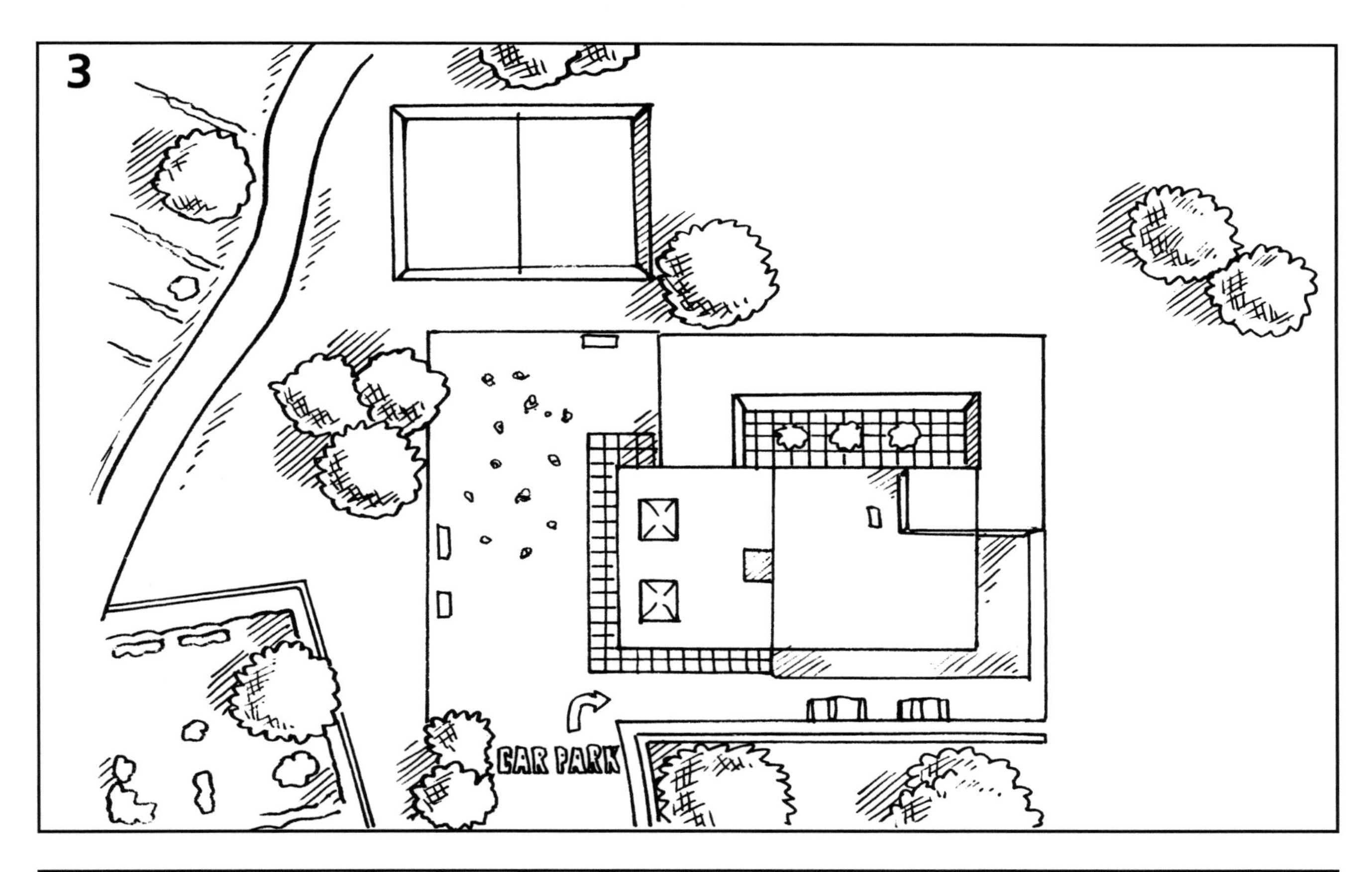

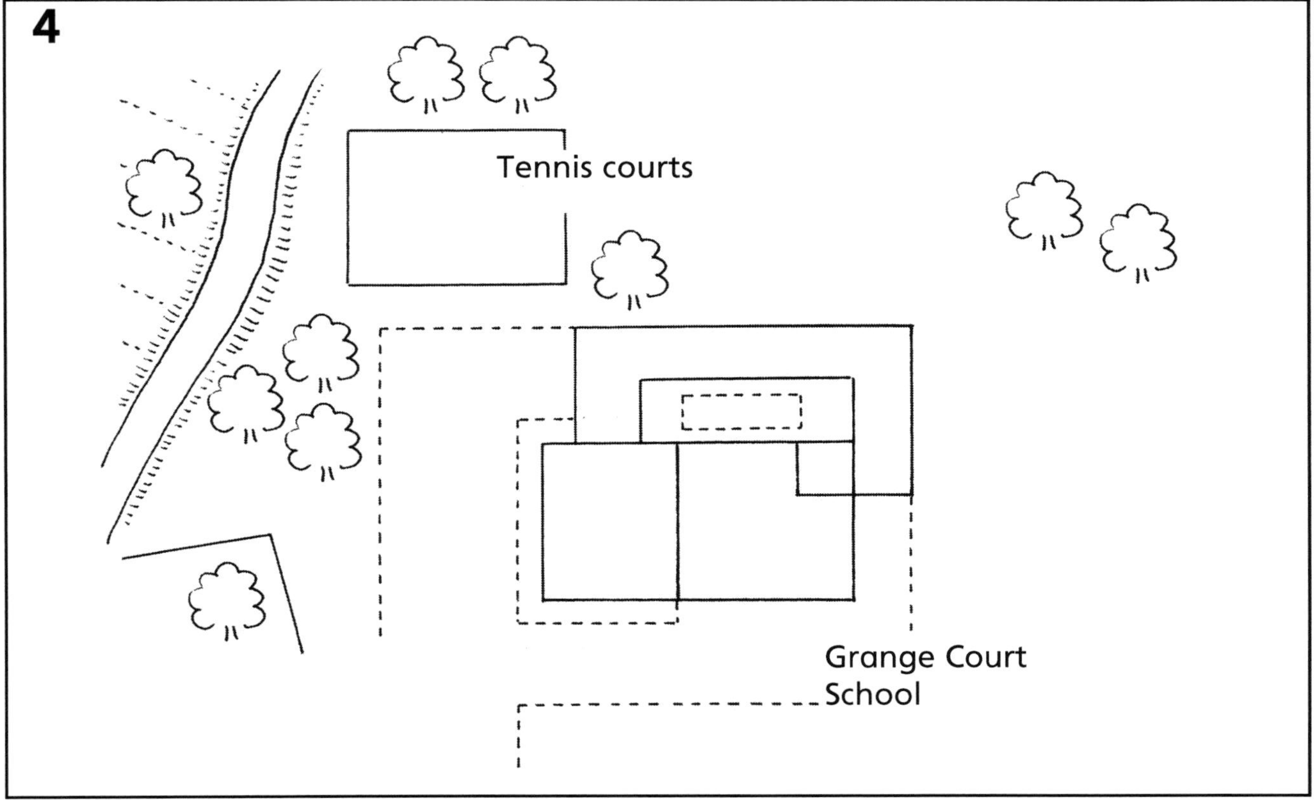

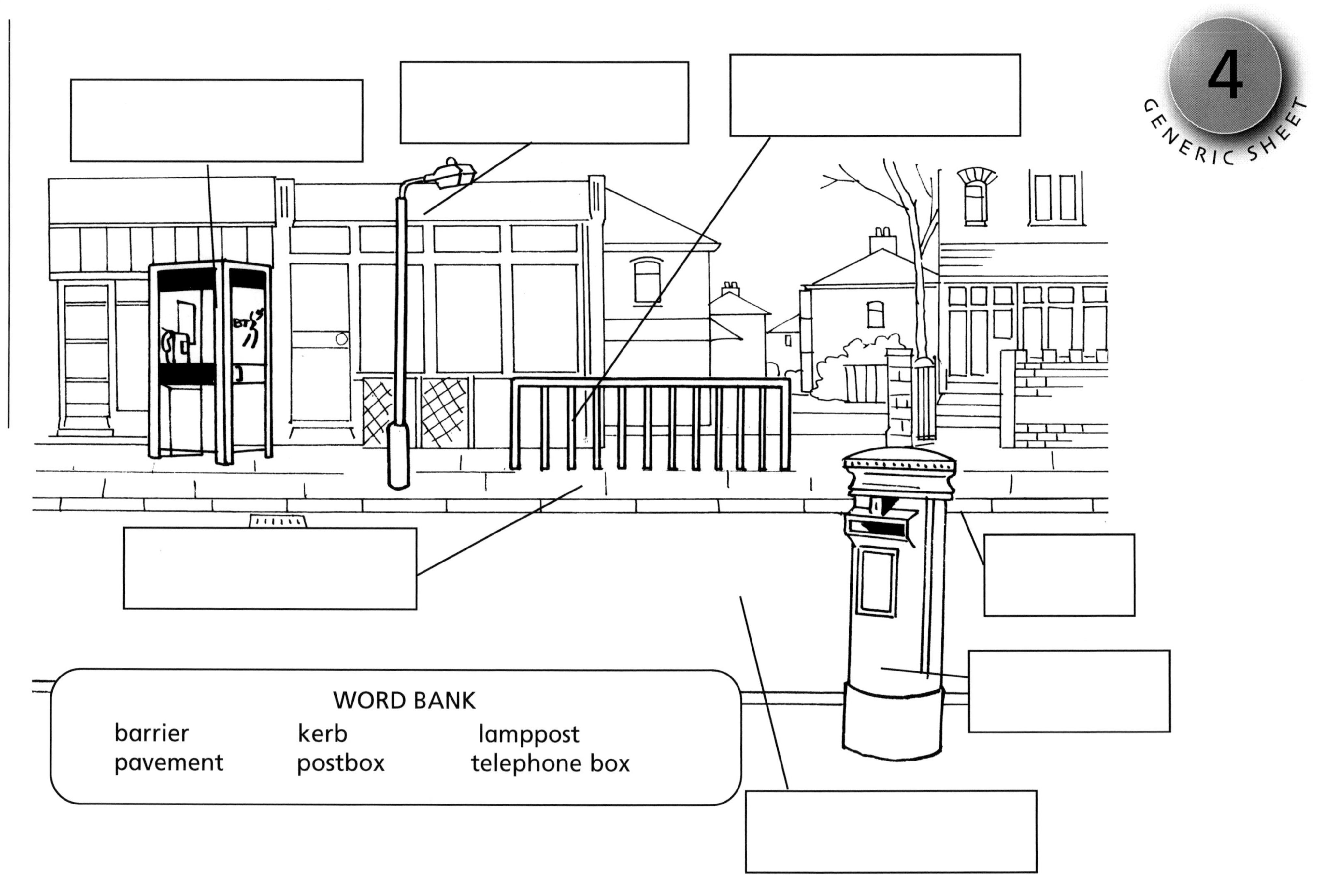
4
GENERIC SHEET
WORD BANK
barrier
pavement
kerb
postbox
lamppost
telephone box

Parking

Name ___________________________________

Parking and safety

This street shows four ways to control parking and keep children safe.

Cut out the labels at the bottom of the sheet and stick them in the correct boxes.

Colour in the yellow lines on the street.

On the back of this sheet, design a sign that says 'Do not cross the road between parked cars'.

a clearway road sign	a pedestrian crossing	a crossing warden	no parking lines

Name ________________________________

Parking and safety

Parking and safety is controlled by:

- pedestrian crossings
- yellow lines meaning 'no parking'
- crossing wardens to help people to cross
- urban clearway signs meaning 'no parking'.

On the map draw:

- a pedestrian crossing
- a warden helping two children to cross the road
- places where yellow lines should be used
- a place for a clearway sign.

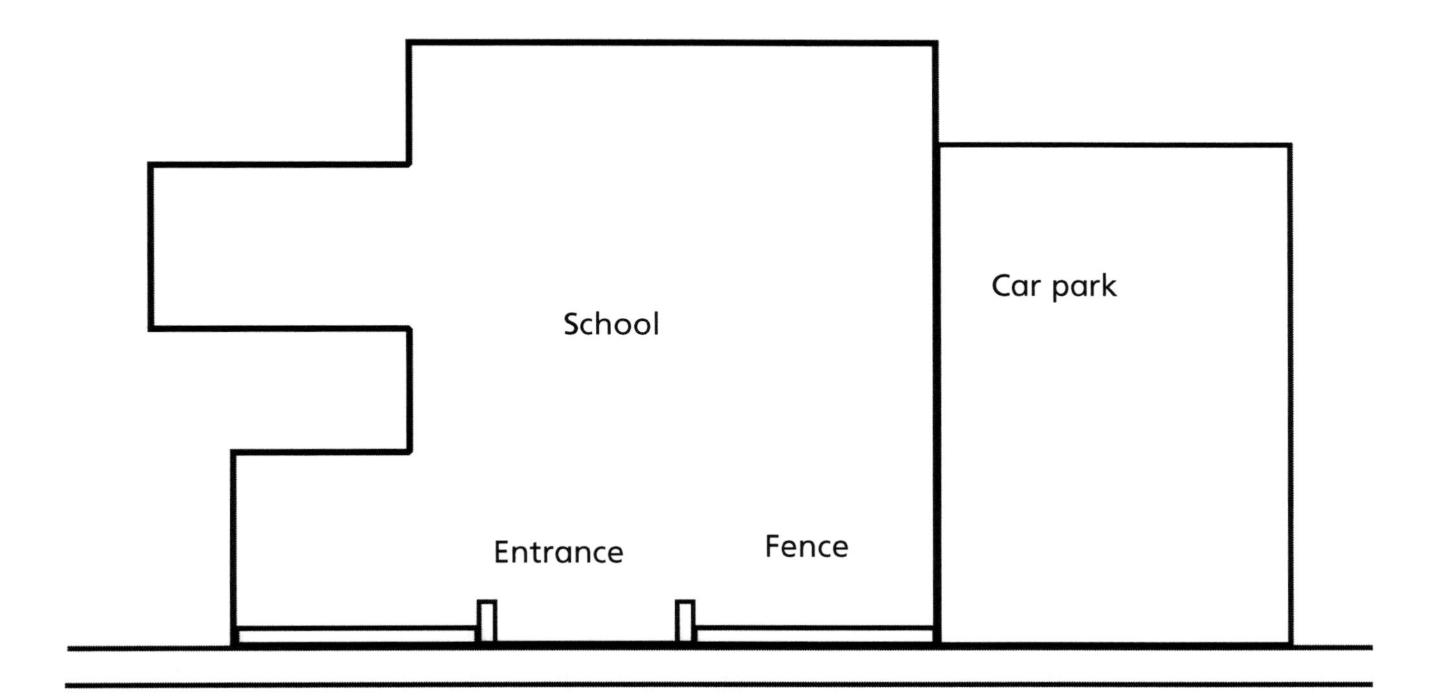

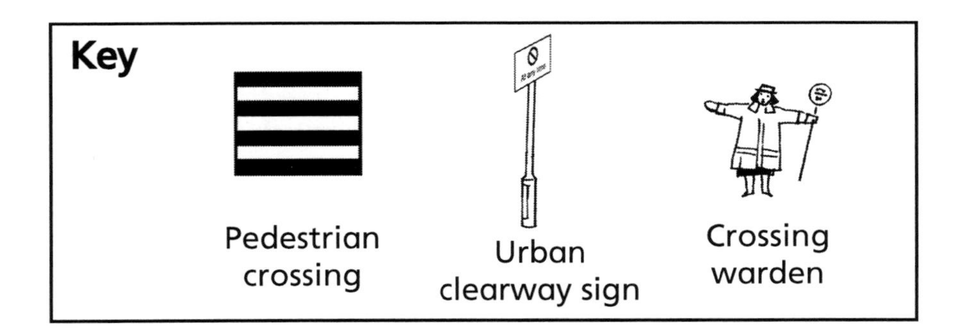

Name ______________________________

Parking and safety

The council wants to build a pedestrian crossing here to help children get safely to school. Where should it go? Draw it on the picture and on the map. On the back of this sheet, write a sentence explaining why you chose this place.

Also on your picture and your map:

- draw a place for a crossing warden and design a symbol to go on the map
- draw yellow lines to stop parking
- draw a place for a clearway sign that drivers can easily see.

On the back of this sheet, write a sentence explaining why you chose these places.

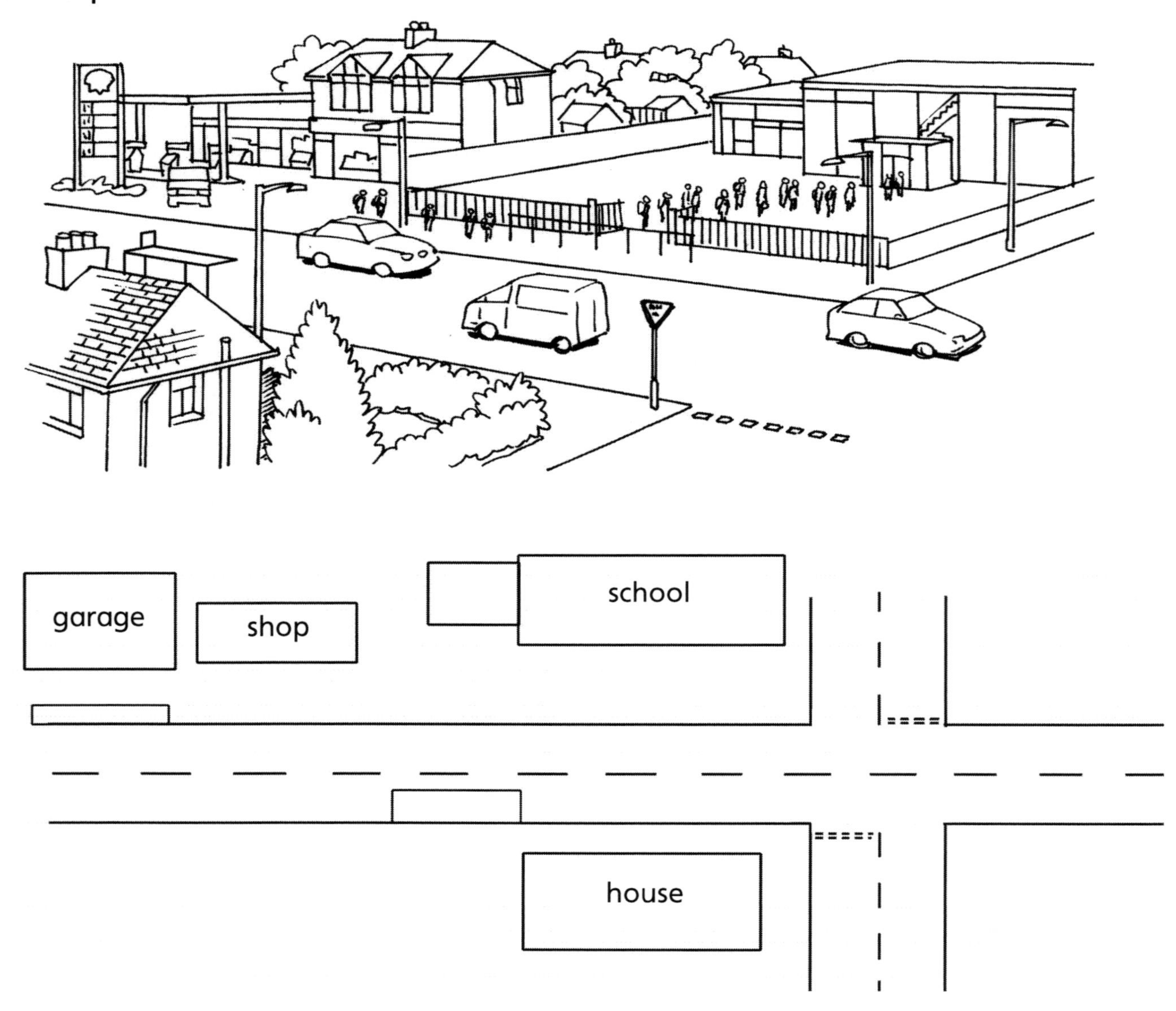

Danger and safety

TEACHERS' NOTES

There is a range of factors to consider with reference to children's safety in general, and on the roads in particular. We all need to be aware of these factors and take them into consideration because many people still fail to take account of children's special and unique problems.

Physically

Children are at greater risk from accidents because their perception of the environment is limited by their size, age and inexperience and because they may often be without adult supervision. Their size restricts their view of their surroundings and this means they have a limited view of traffic. Their size also means that they may not be visible to motorists. The visual systems of young children are also not completely developed until they are 11 or 12 years old, so they may take longer to focus than adults and their peripheral vision is limited. In a noisy environment, such as a busy main road, young children also have difficulty in picking up language clues.

Perception

Children are frequently absorbed by their own interests and this may make them oblivious to their surroundings. In particular, they have difficulty in judging the speed of vehicles and are only able to concentrate on one feature, such as size, at a time in potentially dangerous traffic situations. In their early years, children find it difficult to think through something of which they have no personal experience, so they need help and practice in dealing with dangerous situations such as crossing busy roads. If they have not met a situation before, they are not always able to predict the likely consequences. Their memories tend to be very short-term, so they may forget to look around carefully before trying to cross a road. They also confuse sound and safety. For example, they tend to see a noisy motorcycle as a threat but regard a quieter motor car as less dangerous.

Behaviour

Many children have false ideas and misconceptions that emerge when they are faced with a difficult situation. For example, many children think that the safest way to cross a road is to run as quickly as possible. Running out into the road is one of the commonest ways in which children are injured. Some children also think that because they know that zebra crossings are safe, they can walk on to them at will and that cars will be able to stop immediately. Hence children may not wait for traffic to stop before they step out. Road and traffic signs do not mean a great deal to children, or they may be totally misunderstood. Another problem is that when children are together they often tend to cross the road as a group, and do not watch the road as individuals. They need help in learning not to rely on others for their own safety.

Danger and safety

Geography objectives (Unit 2)
- To express views about making an area safer.
- To recognise ways of changing the environment.

Resources

- Generic sheets 1 and 2 (pages 114 and 115)
- Marker pen
- Sticky tape
- Activity sheets 1–3 (pages 116–118)

Starting points: *whole class*

Explain to the children that they are going to look at dangers in the environment and at what can be done to prevent them. Show them Generic sheet 1 on an OHP or in an enlarged version. Tell them that there are at least five things in the picture that could be dangerous or unhealthy and ask them to identify each one. (Books on top of cupboard could fall; frayed cable on computer could cause fire; pushed over chair and bag could be tripped over; child leaning out of the window; rubbish under the sink could be slippery.)

As the children point out the dangers, number them on the picture. Ask questions such as:

- Why is this dangerous?
- What might happen?

Write two headings – 'Dangers caused by carelessness' and 'Dangers when things go wrong' – on the board or OHP. Ask the children to direct you in putting each number under one of the two headings.

Next show Generic sheet 2 on an OHP or in an enlarged version. Explain that this shows a street where all sorts of things could go wrong and where there are dangers. Ask the children to identify all the dangers they can see and number them. Again, ask the questions:

- Why is it dangerous?
- What might happen?

Then ask:

- What things should not be happening in the street? (Playing games, dog running loose, broken bottle, digging up road with no barriers.)
- How could the street be made safer? (Railings to separate people and traffic, bollards to stop people parking on the pavement, protection from roadworks, stopping some activities on the street, double yellow lines, cycle lanes and pedestrianisation in some areas.)

Tell the children that they are going to look at examples of dangers in the area. They have to identify the dangers, say what might happen and say what could be done to make the area safer.

Group activities

Activity sheet 1
This sheet is aimed at children who need more support. They are able to identify the main dangers in a street (children chasing ball into road, child crossing behind parked car, dog not on lead). They have to spot the dangers and explain what might happen. Then they have to draw what will happen in the same scene a minute later, if the dangers are not addressed.

Activity sheet 2
This sheet is aimed at children who can work independently. They are able to spot a wide variety of dangers (cycling on pavement, running across a zebra crossing without looking, overloaded cycle basket, crossing road carrying planks on shoulder so can't see one way, waving while cycling, chasing ball into the road, dog loose, changing tyre on pavement, van parked on pavement, van doors left open, child playing on corner behind van). They have to identify and number the dangers and explain the risks. Then they have to describe what could be done to address each danger.

Activity sheet 3
This sheet is aimed at more able children. They are able to identify a range of dangers (playing football in the street, vehicle exhaust fumes, cracked paving slabs, unguarded hole in pavement, dog not on lead, sagging overhead cables). They have to explain the nature of the dangers and suggest what could be done to prevent them.

Plenary session

Share the outcomes of the activity sheets. Focus on the range of actions that can make an area safer:

- Pedestrianisation – separating people and traffic.
- Cycle lanes – to separate cyclists from other road users and keep them off the pavement.
- Fences or railings – to keep people on the pavement.
- Bollards to stop parking on the pavement.
- Pedestrian crossings.
- Being on the lookout for dangers in the environment.

Ideas for support

The children spot the issue of the football bouncing into the road quite quickly. Talk about why this is dangerous for the children and the motorcyclist. Then talk about the child crossing from behind the car. Stress that drivers may not see him until he is out into the road and they may not be able to stop in time. The third issue of the dog may again need some discussion. Talk about how the dog might run out into the road if startled by the sound of the motorcycle and discuss what that might lead to.

Take the children for a walk around the school. Tell them that they have to be on the lookout for dangers. These might include:

- water spilled on the floor;
- broken floor tiles;
- bags and coats left on the floor in rooms or corridors;
- narrow corridors, which might become jammed at lunch time;
- broken windows.

Ideas for extension

Ask the children to design and paint signs to be put up around the school warning others of the dangers they have seen. These signs could be in the form of road signs. Stress the need for pictorial signs without words.

Take the children for a walk around the local area. Give each child a map of the area and ask them to mark in the dangerous places as you walk. They can decide on signs for the different dangers and use these in a key. They can also take a camera and record each danger. Use the photographs to form part of a display, together with the map of 'Dangers in our area'.

Typical dangers that are spotted on the walk might be:

- cars parked on double yellow lines or on pavements;
- glass and rubbish in the street;
- gamaged pavements;
- lack of a pedestrian crossing;
- lack of fencing to separate children from traffic.

Linked ICT activities

Using a digital or conventional camera, take photographs of dangerous places around the school environment. Show the children the photographs – on the computer screen if possible (a whiteboard would be really useful for this activity). Ask them to identify what might be dangerous for them in the picture – for example, broken glass, litter, a busy street, a broken paving slab, a hole in the road and so on.

Using the photographs or printouts of the digital images, place the images face down on a table and ask the children to choose one of the pictures at random. Using the image as a starting point, they should identify three things in the picture that they think may present a danger to themselves or other people and animals. They should write three sentences describing what the danger could be and three sentences to say how the place could be made into a safe place. Use a simple word processing program such as *Talking Write Away* or *Textease* to create the sentences. It may also be useful to provide the children with a word bank with useful 'safe' and 'dangerous' words to help them to write their sentences.

Danger and safety

GENERIC SHEET 1

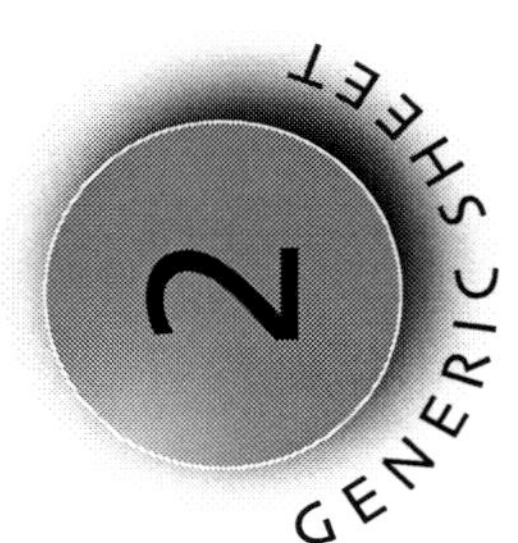
2
GENERIC SHEET

POST OFFICE

Name ______________________________

Danger and safety

There are dangers in this street.
Draw a circle around each danger and give it a number.
Write a sentence about the danger next to the number.

1 ______________________________

2 ______________________________

3 ______________________________

On the back of this sheet, draw this street one minute later.

Name ___________________________

Danger and safety

There are lots of dangers in the street below.
Put a number by each danger you spot.
On the back of this sheet, write a sentence about each danger.
Write another sentence saying how to prevent each danger.

Name ______________________________

Danger and safety

There are at least six things in the picture that could be a danger.
Give each one a number.
In the table below, write a sentence on what the danger is and explain what could happen. Continue on the back of this sheet.
What could be done to prevent each of the dangers you have seen?

Danger	What might happen

Useful resources

Resources recommended for linked ICT activities

Software

Counter for Windows, Granada Learning/ BlackCat Software, Granada Television, Quay Street, Manchester M60 9EA, Tel 0161 827 2927 www.granada-learning.com

My World, Granada Learning/SEMERC, Granada Television, Quay Street, Manchester M60 9EA, Tel 0161 827 2719 www.shop.granada-learning.com/bin/venda

Textease 2000, Softease and Textease 2000 (PC), Softease Ltd, Market Place, Ashbourne, Derbyshire DE6 1ES, Tel 01335 343 421, Fax 01335 343 422 www.softease.com/textease.htm

Talking Write Away, Granada Learning/ BlackCat Software, Granada Television, Quay Street, Manchester M60 9EA, Tel 0161 827 2927 www.granada-learning.com

Dazzle, Granada Learning, Granada Television, Quay Street, Manchester M60 9EA, Tel 0161 827 2927 www.granada-learning.com

Pick a picture, **Granada Colours,** Granada Learning, Granada Television, Quay Street, Manchester M60 9EA, Tel 0161 827 2927 www.shop.granada-learning.com/bin/venda

Valiant Roamer, Valiant Technology, www.valiant-technology.com/ideas1.htm

Clicker 4, Crick Software Ltd, Tel:-01604 671691 www.cricksoft.com

Useful sources of information about change

- The planning department of the local council will have information and maps on recent and proposed changes to your area.
- The local reference library will have information on the local area in the past, including trade directories and maps.
- Make the most of any renovations and building works in the school to record the 'before', 'during' and 'after' in words and pictures.